Lead Halide Perovskite Solar Cells

David J. Fisher

Copyright © 2020 by the author

Published by **Materials Research Forum LLC**
Millersville, PA 17551, USA

Published as part of the book series
Materials Research Foundations
Volume 75 (2020)
ISSN 2471-8890 (Print)
ISSN 2471-8904 (Online)

Print ISBN 978-1-64490-080-2
ePDF ISBN 978-1-64490-081-9

Distributed worldwide by

Materials Research Forum LLC
105 Springdale Lane
Millersville, PA 17551
USA
http://www.mrforum.com

Printed in the United States of America
10 9 8 7 6 5 4 3 2 1

Table of Contents

Introduction

The field of materials science in the 21st century seems to consist of a never-ending succession of new wonders, and one of these – which was hardly discussed a decade ago – is the present class of material. As well as their use as solar cells, they find a place in applications such as photo-detection, light-emitting diodes and lasers; and this versatility can be traced to their unique optical, electrical and crystalline characteristics.

Metal halide perovskites in the form of nanocrystals are particularly useful, due to their very high luminescence efficiencies and wide range of luminescence wavelengths. A combination of single-dot spectroscopy, photon correlation spectroscopy, femtosecond transient absorption spectroscopy and time-resolved photoluminescence spectroscopy reveals the dynamics of excitons, trions and bi-excitons in such nanocrystals. The understanding of trion dynamics is thought to be especially important because they govern the luminescence and are related to ionization processes in the material.

One notable advantage, with regard to their preparation, is that the perovskites can be grown from solution; a relatively low-energy crystallization route. On the other hand, it is somewhat ironic that materials which possess such promising opto-electronic properties should be unstable, and badly degrade in environments which are subject to lengthy and continuous illumination. Much of the research into the material has thus centred upon improving its photo-stability by closely examining the underlying mechanisms. These include photo-induced dissociation, phase separation and phase transformation. These in turn involve the nature of the defect states, thermodynamic factors and details of the chemical bonding.

The solar-cell possibilities of the material are naturally to the fore in its development due to ongoing concerns with regard to the twin problems of a continuous decrease in fossil-fuel resources and the simultaneous risk of global-warming due to the use of those same resources. More efficient conversion of sunlight into electrical energy would help to alleviate both problems, especially if that efficiency could eventually overtake that of silicon solar cells.

Mixed organic–inorganic perovskite solar cells have been shown to offer the combined advantages of low-cost preparation and a high power-conversion efficiency, which has already been strikingly increased during just the first few years of development (figure 1)[1]; that is, from 3.8 to 22.1% between 2009 and 2016. A linear extrapolation of the points of the graph suggests that the maximum value would be over 25 in 2019 and, in fact, a value of 25.2 has been reported.

Materials Research Forum LLC
https://doi.org/10.21741/9781644900819

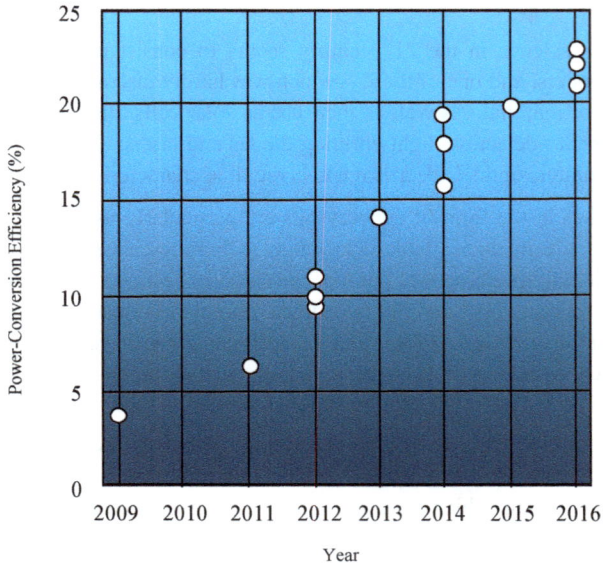

Figure 1. The rapid improvement in the power-conversion efficiency of perovskite solar cells

It is to be recalled that perovskites have the generic formula, ABX_3, where A is a monovalent ion and B is a divalent ion; with X being oxygen, carbon, nitrogen or a halogen. Corner-linked octahedral BX_6 units form a 3-dimensional cubic structure which can be classified as being of calcium-titanate type.

Some useful incidental features of the perovskites are their ability to form high-quality semiconducting films via solution-processing methods, and their tendency to exhibit reversible phase transformations as a function of temperature. That is, their stable orthorhombic phase is found at 100K; with a transition between the tetragonal phase and an orthorhombic phase occurring at about 160K. Grain boundaries moreover have no appreciable effect upon the electrical properties, and they have a band-gap which leads to a high absorbance, while a balanced electron–hole transport behaviour leads to good charge carrier mobilities. They also exhibit a high absorption coefficient and a long exciton-diffusion length; characteristics which are promising for photovoltaic

4

applications. In the specific case of solar-cell applications, they offer proper electronic band alignments, a high level of light absorption, bipolar conductivity, controlled doping possibilities and ferroelectric properties. There are other, more practical, considerations to be taken into account; such as cost, stability, current-voltage hysteresis and a toxicity due to the lead content.

This has prompted a search for lead-free perovskites, even though theoretical knowledge concerning the electronic and defect properties of lead and lead-free halide perovskites suggests that lead halide perovskites will inevitably exhibit better photovoltaic properties than will lead-free perovskites. It is useful to look at some of the drawbacks of lead-free alternatives before returning to the main theme of this work.

In order to aid the search for new halide perovskites, attention has been drawn to the value of the ionic radius as a guide to their prediction and design. The library of Shannon radii-data has recently[2] been more than doubled in size by extending it to include unusual oxidation states and arbitrary coordinations. A new criterion has meanwhile been proposed for predicting the formation of perovskites, and general predictive criteria involving radii and normalized-volume data have been discovered for predicting the band-gaps of a wide range of 3-dimensional halide perovskites.

The perovskite structure is rather amenable to replacement and many elements, including cobalt, europium, germanium, iron, manganese, palladium and tin can be incorporated; with the last 2 elements in particular being proposed as alternatives to lead. Other alternative substituents include organic cations such as formamidinium, and the introduction of long-chain organic groups at the A-site of ABX_3 perovskite-type layer compounds is possible. It has been shown[3] that the superior photovoltaic properties of the lead halide perovskites are due to a combination of their high O_h symmetry, the existence of the lone-pair lead 6s and inactive lead 6p orbitals, spin-orbit coupling and the ionic nature of the halides. The replacement of lead by heterovalent non-toxic elements such as In^I, Sb^{III} and Bi^{III} or isovalent elements such as Ge^{II} or Sn^{II} retains the electronic features of lone-pair and inactive orbitals. Here In^I is rather unstable, due to the high-energy position of the indium 5s states. Replacement with Sb^{III} or Bi^{III} leads to alternatives which exhibit low structural or electronic dimensionality. This markedly limits the performance characteristics of the resultant solar cell. The replacement of lead by Sn^{II} or Ge^{II} leads to perovskites which retain 3-dimensional structural and electronic dimensionality, although the former is a better choice due to its lower cost and greater stability. Thus far, tin perovskite solar cells have given the best results among all of the lead-free halide perovskite solar cells. It is noted that Sn^{II} is less stable than is Pb^{II}, due to the fact that the tin 5s orbital is higher in energy than is the lead 6s orbital.

The extremely high optical absorption coefficient and very long carrier diffusion lengths of $CH_3NH_3PbI_3$ perovskites were analyzed[4] using first-principles theory. It was again deduced that the superior properties are due to a combination of direct band-gap p-p transitions, enabled by lead lone-pair s-orbitals and strong antibonding coupling between lead lone-pair s-and iodine p-orbitals; not to mention the perovskite symmetry, high iconicity and large lattice constant. It was shown that the material exhibits intrinsic ambipolar self-doping, with the conductivity being variable from p-type to n-type by controlling the growth conditions. The p-type conductivity can be further improved by incorporating group-IA, -IB or -VIA elements under iodine-rich/lead-poor growth conditions. The n-type conductivity could not be similarly improved, due to compensation arising from intrinsic point defects.

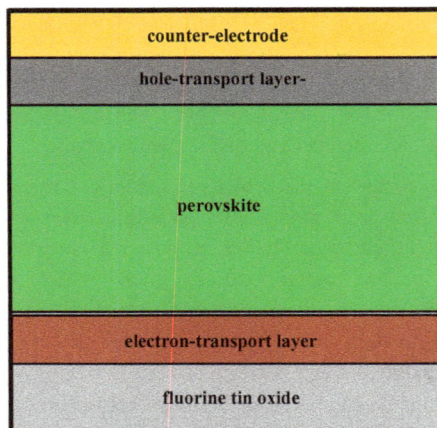

Figure 2. Generic layout of a lead-halide perovskite solar cell

Double-halide perovskites of the form, $A_2B'B''X_6$, where A is cesium or CH_3NH_3, B' is bismuth or antimony, B'' is copper or silver and X is chlorine, bromine or iodine, (tables 1 to 3) have also been investigated[5] as possible alternatives to lead halide perovskites. Most of these double perovskites exhibit a better stability than those of hybrid lead halide perovskites, and this is attributed to replacement of the organic cation by the inorganic cation, Cs^+. They nevertheless still require encapsulation in order to prevent degradation. Most of the halide double perovskites considered have indirect and wide band-gaps which are unsuitable for solar-cell applications, although $Cs_2InAgCl_6$ for example has a direct band-gap that permits its use in solar cells. Only $Cs_2AgBiBr_6$ appears to have been used as a light-absorber in planar heterojunction solar cells, with a power-conversion efficiency of only about 1%. It has been proposed that the inclusion of trivalent cations such as lanthanum could have beneficial effects.

The stability of the 2+ ionic state decreases in going from lead, via tin, to germanium. This constitutes a serious problem. Under dry ambient conditions, stable device operation for up to 500h has been reported, but humidity is a major menace to successful perovskite use. Substituent-alteration has been found to be a possible solution, with an increased stability under humid conditions resulting from the incorporation of a modest proportion of bromine; albeit at the cost of a slight decline in device performance. Encapsulation of the material in water-proof materials nevertheless remains necessary.

Table 1. Calculated formation enthalpies
of double halide perovskites

Perovskite	ΔH (eV)
$Cs_2AgBiCl_6$	+0.57
$Cs_2AgBiBr_6$	+0.38
Cs_2AgBiI_6	-0.41
$Cs_2InBiCl_6$	+0.01
$Cs_2InBiBr_6$	-0.04
$Cs_2NaSbCl_6$	-0.19
$Cs_2NaBiCl_6$	-0.43
Cs_2NaSbI_6	+0.49
Cs_2NaBiI_6	+0.40
$Cs_2NaBiBr_6$	+0.01
$Cs_2NaSbBr_6$	+0.11

The poor operational stability of these devices is an ongoing problem. In order to quantify the degree of progress, a so-called T80 parameter has been used[6]. This was defined to be time required for a solar cell to decay by 20% from its initial power-conversion efficiency value. Although a steady increase in the power-conversion efficiency was observed, the T80 parameter did not exhibit any appreciable improvement. In fact, a levelling-off was detected, with the T80 value stalled at about 6000h. As explained in detail below, the structure of a perovskite solar cell consists of a stack of various functional layers. In addition to the perovskite itself, the degradation mechanisms affecting each of those layers clearly need to be addressed in order to improve the overall lifetime of the device. Data on the long-term stability of 404 organo lead halide perovskite cells, extracted from 181 papers, have been analyzed[7] using machine-learning methods in order to determine the effects of material choice, deposition method and storage conditions upon solar-cell stability. For regular cells, mixed cation perovskites, multi-spin coating for one-step deposition, dimethyl formamide plus dimethyl sulfoxide as a precursor solution and chlorobenzene as an anti-solvent had positive effects upon stability. The use of SnO_2 as the electron transport layer and carbon as the back contact was also found to improve stability. Cells stored under low humidity were confirmed to be more stable. Degradation was slightly faster in inverted cells under humid conditions and the use of materials such as mixed-cation perovskites and NiO_x in their usual roles was found to produce more stable cells.

Among the most studied perovskites in the present context are $CH_3NH_3PbI_3$, $CH_3NH_3PbBr_3$ and $CH_3NH_3Pb(I,Cl)_3$. The best solar cells which are manufactured from $CH_3NH_3PbI_3$ typically exhibit band-gaps of 1.50 to 1.55eV. Because the highest occupied molecular orbital position is similar for the above 3 systems, the lowest unoccupied molecular orbital energy is greatly affected by the nature of the halogen which is involved. This consequently allows for the possibility of tailoring both the energy-level and the highest-lowest occupied-unoccupied energy-gap; a feature which is particularly effective in the case of $CH_3NH_3Pb(I,Cl)_3$. Solar cells which are based upon $CH_3NH_3PbI_3$ can exhibit quantum efficiencies of up to 800nm and can access photons in the visible, and part of the infra-red, spectrum. First-principles calculations are effective in predicting their properties, while optical and electronic measurements reveal critical parameters such as the exciton-diffusion length and charge-transfer density. *Ab initio* calculations can predict new families of photovoltaic perovskites. The efficiency limit of perovskite solar cells is about 31%; close to the Shockley–Queisser limit of 33% which is attainable by GaAs solar cells.

Table 2. Measured mechanical properties of perovskites

Perovskite	B (GPa)	H (GPa)	E (GPa)	CTE (m/K)
$Cs_2AgBiBr_6$	27.3	0.67	22.6	27.8
$CsPbBr_3$	15.5	0.34	15.8	37.7
$CH_3NH_3PbI_3$	10.2	0.42	10.4	43.3
$(CH_3NH_3)_2AgBiBr_6$	7.7	0.55	7.9	44

B: bulk modulus, H: hardness, E: Young's modulus, CTE coefficient of thermal expansion

Another source of poor device stability in perovskite solar cells is due to the interfaces. That is, a decrease in the power-conversion efficiency is associated with chemical degradation of metal contacts which is in turn due to ion motion in the perovskite layer.

Perovskite solar cells are essentially free-carrier based devices wherein charge transport and separation occur in the same manner as in heterojunction solar cells. The thermodynamics of thermal equilibrium, and the Anderson model, dictate that when 2 types of semiconductor are in direct contact, their respective free carriers spontaneously migrate towards one another. The migration then equilibrates the Fermi-energy level and gives rise to a charge-depletion region having an associated electric field; thus constituting a junction. Smooth energy-band bending can occur at a p–n junction, but an energy-step always appears at the heterojunction interface due to band-offset. The latter greatly affects the voltage-output of a perovskite solar cell. Interfacial electronics and energy can also directly influence charge extraction and collection. Terahertz time-domain spectroscopy[8] has here proved itself to be a valuable non-contact technique for monitoring the dynamics of carriers, phonons and excitons in this type of material. Non-adiabatic molecular dynamics methods, combined with *ab initio* time-domain density functional theory, have meanwhile permitted the modelling of time-resolved spectroscopy at the atomistic level[9]. The method can treat point defects, surfaces, grain boundaries, dopants and interfaces, and provide important insights into the mechanisms of charge and energy loss.

Table 3. Electronic properties of double perovskites

Perovskite	Space Group	Band-Gap (eV)	PCE (%)
$Cs_2AgInCl_6$	Fm3m	3.23 to 3.3	-
$Cs_2AgBiCl_6$	Fm3m	2.2 to 2.77	-
$Cs_2AgBiBr_6$	Fm3m	1.8 to 2.19	1.22 to 1.44
$Cs_2Au^IAu^{III}Cl_6$	I4/mmm	2.04	-
$Cs_2Au^IAu^{III}Br_6$	I4/mmm	1.31	-
$Cs_2Au^IAu^{III}I_6$	I4/mmm	1.60	-
$(CH_3NH_3)_2AgBiBr_6$	Fm3m	2.0	-
$(CH_3NH_3)_2KBiCl_6$	R3m	3.04	-
$(CH_3NH_3)_2AgSbI_6$	R3m	1.93	-

PCE: power-conversion efficiency

A typical perovskite solar cell consists of a transparent substrate, a perovskite layer situated between electron-transport and hole-transport layers, and a metallic upper counter-electrode. The transparent substrate is usually made from fluorine-doped or indium-doped tin oxide. In mesoscopic and n–i–p type perovskite solar cells, illumination first impinges on the electron-transport layer while, in inverted planar heterojunction structures, illumination first impinges on the hole-transport layer. In mesoscopic and planar heterojunction cells therefore, the interfaces of the cells are mainly of substrate|electron-transport-layer, electron-transport-layer|perovskite, perovskite|hole-transport-layer and hole-transport-layer|counter-electrode type. In the case of inverted planar heterojunction solar cells, they are of substrate|hole-transport-layer, hole-transport-layer|perovskite, perovskite|electron-transport-layer and electron-transport layer|counter-electrode type.

Most perovskite solar cells have gold or silver counter-electrodes which are deposited via vacuum evaporation or electron-beam sputtering, although aluminium, molybdenum, nickel and silver-aluminium alloys have also been used for that purpose. Carbon-based materials have also been considered to be very suitable for use as counter-electrodes, as they offer easy preparation, good chemical stability, compatible energy levels and low cost.

During illumination the photons are absorbed by the perovskite, and free electrons and holes are excited. Free carriers then first travel through the perovskite via migration or diffusion and may then recombine or be scattered by defects and crystal boundaries. The electrons and holes are subsequently gathered by the electron-transport and hole-transport layers, respectively. In order to ensure a high power-conversion efficiency, all of the photo-induced carriers should reach the external load, but various defects arising from lattice mismatch, energy-level misalignment and thermal noise are always present in the interfaces. The electron-transport layer prevents the holes which are generated in the perovskite from reaching the cathode. The latter can also be considered to be a hole-blocking layer.

When the lowest unoccupied molecular orbital of the electron-transport layer is far lower than that of the absorber, there is a decrease in the open-circuit voltage. The highest occupied molecular orbital of the hole-transport layer should be just slightly higher than that of the perovskite, in order for it to be possible to collect holes and block electrons. The electron-transport-layer|perovskite interface thus plays an essential role in determining the open-circuit voltage. Meanwhile the perovskite|hole-transport-layer interface is more important to control of the photo-current. The work functions of adjacent layers in perovskite solar cells, which should be closely matched, can however be adjusted by means of interface engineering.

It is the high open-circuit voltage, together with slow recombination, that makes lead-halide perovskites such promising photovoltaic materials. The microscopic basis of charge-carrier lifetimes is particularly interesting, as the latter depend upon defect positions, defect densities and the kinetic factors which control phonon-assisted interactions between extended states, in the conduction and valence bands, and localized defect-states. It is notable that, while thin films of these perovskites are not defect-free, few of the possible intrinsic defects have energies which are close to the mid-gap. This renders them remarkably recombination-active. Many shallow defects may nevertheless be present, due to the existence of off-stoichiometric conditions or strain during film-formation. Shallow defects can trap electrons or holes efficiently, but generally not both. Transients in experimental measurements, such as photoluminescence enhancement and hysteresis suggest that charge-trapping ionic vacancies and interstitials may self-destruct upon illumination. It is therefore difficult to determine the defect energy-levels, densities and even type. Defects can also make the material p-type or n-type.

Table 4. Structures of methylammonium lead trihalides

Halide	Temperature (K)	Structure	a(Å)	b(Å)	c(Å)
Br	>236.9	cubic (Pm3m)	5.901		
Br	155.1 to 236.9	tetragonal (I4/mcm)	8.322	11.832	
Br	149.5 to 155.1	tetragonal (P4/mmm)	5.894	5.861	
Br	<144.5	orthorhombic (Pna2$_1$)	7.979	8.580	11.849
Cl	>178.8	cubic (Pm3m)	5.675		
Cl	172.9 to 178.8	tetragonal (P4/mmm)	5.656	5.630	
Cl	<172.9	orthorhombic (P222$_1$)	5.673	5.628	11.182
I	>327.4	cubic (Pm3m)	6.329		
I	162.2 to 327.4	tetragonal (I4/mcm)	8.855	12.659	
I	<162.2	orthorhombic (Pna2$_1$)	8.861	8.581	12.620

The defect chemistry of these perovskites has been investigated mainly theoretically, using density functional theory, but the complexity of the electronic structures of the materials makes the accuracy of such calculations somewhat questionable. In the particular case of the compound, $CH_3NH_3PbI_3$, the effect of the exchange-correlation functional upon the electronic structure and defect formation energies was determined[10] by comparing semi-local and hybrid functionals; with and without spin-orbit coupling corrections. The results showed that the defect chemistry of $CH_3NH_3PbI_3$ is governed by normal iodine redox behaviour, as demonstrated by the defect properties of interstitial iodine. Quantitative differences were expected to occur upon changing the halide, but without greatly affecting the underlying chemistry. An absence of any marked directional preference of lead halide bonds made lead-based defects less relevant to trapping effects.

This particular perovskite (table 4)[11] has proved to be especially problematic because it appears to be thermally unstable when in an inert environment, and tends to decompose into CH_3NH_3I and PbI_2 at just slightly elevated temperatures. Even the most stable perovskite solar cells currently survive annealing for up to only 500h at 85C and clearly cannot yet compete with rival materials, such as silicon, cadmium telluride and gallium arsenide, with their guaranteed lifetimes of up to 25 years.

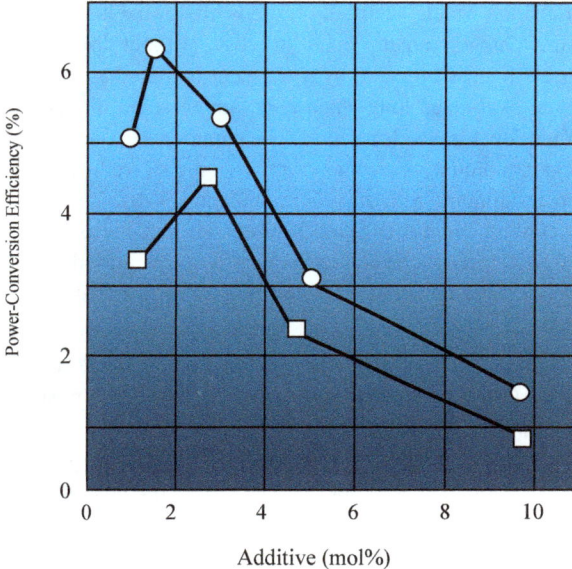

Figure 3. Power-conversion efficiency of $CH(NH_2)_2SnI_3$ as a function of additive content. Upper curve: KHQSA (hydroquinone sulfonate potassium salt), lower curve: APSA (2-aminophenol-4-sulfonic acid)

The grain boundaries and defects cause detrimental recombination loss, and a reduction in the power-conversion efficiency of solar cells made from lead-halide perovskite films. The grain boundaries in such films are in fact far less harmful than are those in polycrystalline solar cells which are made from silicon. Ion migration in polycrystalline perovskites is dominated by its grain boundaries. The well-known problematic penetration of moisture into perovskite films arises from the diffusion of water molecules along the grain boundaries. It is therefore important to improve the quality of these films and thus their opto-electronic performance. One strategy is to promote homogeneous nucleation at the surface of a previously-formed layer so as to obtain a smooth perovskite film with a high surface coverage.

Another route is that of compositional tailoring. The unit cell of α-phase lead-halide perovskites, ABX_3, contains 5 atoms which are arranged in a cubic structure, and the

$CH_3NH_3PbI_3$ perovskite exhibits a reversible cubic-to-tetragonal (β-phase) transition at about 56C; a transition which is partially responsible for the instability of that perovskite. In order to ensure a high-symmetry cubic structure, without distortion, the A-cation must fit into the space which is bounded by 4 adjacent corner-sharing BX_6 octahedra. When the A-cation is an over-sized long-chain alkyl amine anion, the perovskite takes on the form of a 2-dimensional layered structure. The commonest A-site cation substitutions are CH_3NH_3+methylammonium, with an ionic radius of 1.8Å and $HC(NH_2)_2+$formamidinium, with an ionic radius or 1.9 to 2.2Å. Isopropanol containing phenylethylammonium iodide or n-butylammonium can be used to convert the upper layer of $CH_3NH_3PbI_3$ into a layered structure. It is obvious that the huge number of organic cations offers innumerable possible additives for the fine-tuning of interfaces.

The forces, such as hydrogen bonding, which are created by additives between the cations and anions of perovskite precursors can lead to the formation of intermediate phases and thence to the appearance of high-quality films. Compact uniform perovskite films can both increase the power-conversion efficiency of devices and improve their stability under ambient conditions. For example, the molecular structure of some typical additives such as 2-aminophenol-4-sulfonic acid and hydroquinone sulfonate potassium salt (figure 3) leads[12] to interaction between the sulfonate group and Sn^{2+} ions, thus causing the perovskite particles and an $SnCl_2$ additive composite layer to be encapsulated *in situ*.

Cesium is a popular addition to inorganic lead halide perovskites because of its potential ability to improve their thermal stability, even at the risk of provoking phase-instability[13]. The opto-electronically active black phases, α-, β-, and γ-$CsPbI_3$, are metastable at room temperature, and easily transform into an opto-electronically inactive yellow phase. Crystals of $CsPbX_3$, where X is bromine, chlorine or iodine make very good room-temperature X-ray and γ-ray detectors[14]. They are also related to the so-called zero-dimensional Cs_4PbX_6 materials, again with X being bromine, chlorine or iodine. Samples can be divided into 3 groups: pure Cs_4PbBr_6, Cs_4PbBr_6 with defects and Cs_4PbBr_6 with 3-dimensional inclusions.

An all-inorganic $CsPbBr_3$ perovskite solar cell is a promising solution to the problem of the high efficiency yet poor stability of organic–inorganic cells. Putting inorganic hole-transport layers at the perovskite/electrode interface decreases charge-carrier recombination without compromising the higher resistance to environmental attack. Substituted p-type inorganic $Cu(Cr,M)O_2$ (M = Ba, Ca or Ni) nanocrystals, having higher hole-transporting characteristics due to increased interstitial oxygen, effectively extract holes from perovskite. All-inorganic $CsPbBr_3$ devices having the structure, fluorine-tin-oxide|TiO_2|$CsPbBr_3$|$Cu(Cr,M)O_2$|carbon can attain[15] an efficiency of up to 10.18%, which

Materials Research Forum LLC
https://doi.org/10.21741/9781644900819

further increases to 10.79% upon doping samarium ions into the perovskite halide. A non-encapsulated $Cu(Cr,Ba)O_2$-based cell exhibits moreover a marked stability in air with 80% relative humidity for 60 days, at 80C for 40 days or under illumination for 7 days.

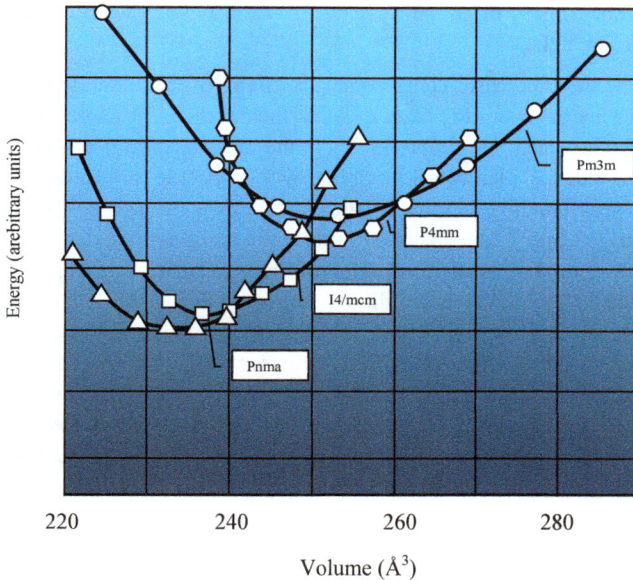

*Figure 4. Calculated energy barriers for transitions between
the tetragonal and orthorhombic phases of $CH_3NH_3PbX_3$*

An interesting reversible phase transformation between Cs_4PbBr_6 and $CsPbBr_3$, uncommon in inorganic materials, permits the possible recycling of perovskite materials[16]. An advantage of such a transformation is that it is possible to produce nanomaterials exhibiting a wide range of optical responses: for example, a rapid transformation from colourless 0-dimensional material to a highly-coloured 3-dimensional material. Hybrid structures of the form, $CsPbBr_3|Cs_4PbBr_6|Rb_4PbBr_6$, can exhibit photoluminescence quantum yields of 70 to 90%; several orders-of-magnitude greater than that of plain $CsPbBr_3$. A shell of 0-dimensional perovskite which is just a few nanometres thick can reduce the insulation of a 3-dimensional perovskite core.

Assemblies of the form, $CsPbBr_3|Cs_4PbBr_6$, can also function as lasers, monocrystalline white LEDs and photodetectors. A systematic theoretical investigation of $CsPbI_3$ perovskite polymorphs has been made[17] by using pseudo-potential plane wave methods and hybrid functionals in order to estimate band-gaps and other properties. The predicted elastic properties reflected the flexibility and softness of these perovskites. Their absorption coefficients of 10^4 to 10^5/cm, and low exciton binding energies, confirmed their suitability as solar cells.

The lead halide perovskites also compete in the field of colloidal semiconductor nanocrystals, where the most striking feature is that their copious structural defects tend not to impair their opto-electronic properties[18]. On the other hand, a persistent problem with $CsPbX_3$ nanocrystals is a structural instability with respect to moisture, oxygen and heat. A relatively weak bonding between ligands and particle surfaces always results in damage to the nanostructure[19]. Oleic acid and oleylamine are the ligands which are the most commonly used for the synthesis of $CsPbX_3$ nanocrystals. Capping-ligands can however easily detach from the surface during treatment, leading to aggregation and structural collapse. Another route to improving stability is passivation by inert shells made from materials such as SiO_x and AlO_x. The nanocrystals can then at least retain their original properties. The coating of $CsPbX_3$ nanocrystals with semiconductors such as ZnS at the single-particle level when preparing heterostructures is a strategy which both protects the nanocrystal and also optimises the electronic structure for various applications. Metal-halide perovskite quantum-dots have also been considered to be promising materials for opto-electronic and other applications[20]. Cesium-lead halide-based perovskite quantum-dots in particular have been potential luminescent alternatives to II-VI- and I-III-VI2-group semiconductor nanoparticles. Their quantum yield of 90% is greater than that of any other semiconductor quantum-dot. The morphology of the material's nanoplates, nanowires, nanocube and nanosheets confirms their self-assembly behaviour in a colloidal medium.

Hybrid organic-inorganic perovskites can be self-assembled by means of chemical deposition of the components, thus integrating the divers properties of the compounds at the molecular scale; even though their various types of interaction can complicate the outcome[21]. The overall electronic properties of the hybrid perovskites, $CH_3NH_3Pb(X,Y)_3$ where X and Y are each bromine, chlorine or iodine, are defined by $Pb-(X,Y)_3$ networks in which the conduction band comprises lead p-orbitals while the valence band is a hybridization of lead s-orbitals and halogen p-orbitals. The organic cation plays more than a structural role in that, although it does not directly contribute to the interacting orbitals, it can markedly affect them. Hydrogen-bonding between the amine group and halide network, and a considerable orientational disorder of the organic cation within the inorganic cage - due to its relatively high rotational mobility – can produce appreciable lattice deformations and disturb the electronic interactions within the Pb−X cage. The resultant structure of the $CH_3NH_3PbX_3$ hybrid is thus a fluctuating one, within which tilting and distortion of the octahedra and rotation and polarization of the dipoles greatly affect the opto-electronic properties. Extending the absorption band to longer wavelengths, and solving the stability problems of metal halide perovskites, have always been primary concerns. Compositional modification offers the possibility of varying the band-gap, and this has led[22] to the development of binary, ternary and two-dimensional/three-dimensional mixed-cation perovskites. Solar cells which are based upon hybrid organic–inorganic lead halide perovskites are thus a topic of great interest[23], given that their energy conversion efficiency can exceed 20%. In order to improve these results even more, anti-reflection coatings have been added at the top substrate surface, optical cavities have been integrated into devices and plasmonic or dielectric nanostructures have been incorporated into the various layers of the cells[24]. As well as their eminent usefulness as solar-cell materials, hybrid lead-halide perovskites are of interest in the fields of light-emitting diodes and lasers[25].

Concerns over the stability problem led to study[26] of the basic chemistry and physics of the crystal structures (figure 4) and of the synthesis of bulk or nanocrystalline specimens. In the case of $CH_3NH_3PbI_3$, the grain size was found to be directly related to the photovoltaic performance of the solar cell[27]. A critical factor which affected the grain size was the temperature at which the $CH_3NH_3PbI_3$ was deposited onto a PbI_2 layer.

Preparation of a thin film of the hybrid perovskite depends upon crystal formation that is regulated by the concentration and ratio of ionic compounds in solution and by the temperature. Crystallization can also proceed via the solid-state reaction of CH_3NH_3I and PbI_2 at about 100C[28]. Experiments in which both the substrate and 0.050M $CH_3NH_3PbI_3$ solution were preheated to -10, 20 or 50C had shown that the grain size depended upon the $CH_3NH_3PbI_3$-solution temperature. Theory suggested that, as the temperature of the solution increased, the equilibrium concentration of the chemical reaction increased due to an increased solubility of the $CH_3NH_3PbI_3$. The critical free energy was higher at high temperatures than at low temperatures, and this determined the number of nuclei; thus resulting in a decrease in the number of nuclei becoming $CH_3NH_3PbI_3$ crystals, and a consequent increase in grain size. The perovskite-layer colour darkened with increasing grain size. The presence of defects in the crystal structure could lead to direct contact between the electron-transport and hole-transport layers. The defects, which were caused by the increase in grain size with temperature, acted as recombination sites between the layers. The absorbance was greatest when the $CH_3NH_3PbI_3$ solution was at 20C, and was lowest when it was at -10C. This decrease was attributed to the decreased grain size.

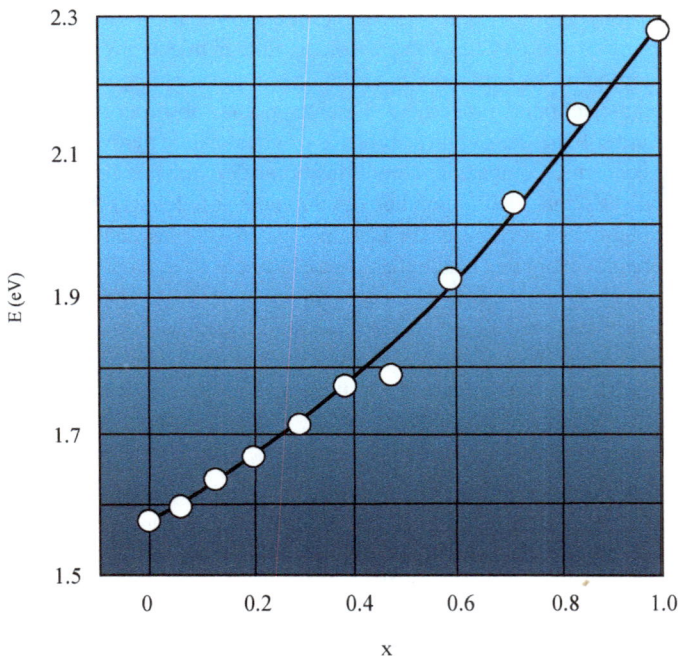

Figure 5. Band-gaps of $CH_3NH_3PbI_{1-x}Br_x$ perovskites

Methylammonium lead iodide is a semiconducting pigment, with a direct band-gap of 1.55eV that corresponds to an absorption onset of 800nm. Its photon-generated excitons have a binding energy of less than 50meV and can thus rapidly dissociate into free carriers at room temperature. The electrons and holes have small effective masses, and this leads to carrier mobilities which range from $24cm^2$/Vs for electrons to $105cm^2$/Vs for holes, with recombination occurring within hundreds of nanoseconds. This in turn leads to carrier-diffusion lengths of 100 to 1000nm and electron–hole diffusion lengths of more than 175μm. The iodine can be replaced at will by bromine or chlorine; although the precise structural location of chlorine has attracted some controversy. The substitution of bromine[29] permits a continuous tuning of the optical band-gap (figure 5). The introduction of bromine also raises the conduction-band and lowers the valence-band. Increasing the conduction-band minimum helps to facilitate energy-band matching

Materials Research Forum LLC
https://doi.org/10.21741/9781644900819

between TiO_2 and $CH_3NH_3PbBr_3$, and improves the open-circuit voltage. The photon-generated excitons of $CH_3NH_3PbBr_3$ have a higher binding energy (150meV) as compared to the 50meV of $CH_3NH_3PbI_3$. The power-conversion efficiency of solar cells based upon $CH_3NH_3PbBr_3$ therefore remains lower than that of cells based upon $CH_3NH_3PbI_3$. Unlike bromine, chlorine encourages the formation of perovskite grains and the increased crystallinity tends to speed up the transfer and diffusion of photon-generated carriers; thus increasing the electron–hole diffusion length. The diffusion length of $CH_3NH_3Pb(I,Cl)_3$ exceeds 1µm; far greater than the 100nm of $CH_3NH_3PbI_3$.

With regard to general perovskite solution-growth, dimethyl sulfoxide, γ-butyrolactone, N-methyl-2-pyrrolidone and N,N-dimethylformamide are suitable solvents for both lead halides and CH_3NH_3I. Complete surface coverage is essential when preparing an interface, in order to avoid recombination-loss, but simple spin-coating proves however to be unsuitable for reliably producing a uniform and homogeneous perovskite layer. Mixtures of the above solvents have however been found to guarantee the formation of very uniform dense perovskite layers.

Traces of PbI_2 which result from the film-preparation process can occupy the perovskite grain boundaries and constitute a blocking layer between a TiO_2 substrate and the perovskite interface. This reduces the number of trapping sites and improves electron transfer.

Electron-Transport Layer

Titanium dioxide is the material which is most frequently used as an inorganic electron-transport material in perovskite solar cells, due to its strength, favorable energy levels and low cost. Electron-injection from the perovskite to a TiO_2 electron-transport layer is very fast, but electron-recombination is extensive because of low electron mobility and poor transport. A relatively high density of trap-states in TiO_2 also reduces solar-cell efficiency and stability. Perovskite solar cells which involve TiO_2 tend to require high-temperature sintering, and this is a drawback at the production stage. Various TiO_2 nanostructures, such as nanowires, nanotubes and nanorods, have nevertheless been considered as being effective in collecting electrons at the interface between the oxide and the perovskite. The use of a bi-layered TiO_2 blocking layer can produce an hierarchical oxide structure and a perovskite solar cell having a power-conversion efficiency which is better than 16%.

The properties of TiO_2 can moreover be modified by doping with aluminium, magnesium, neodymium, niobium, tin, yttrium, zinc and zirconium. Doping with Y^{3+}, for example, has promoted the efficient extraction of photo-generated charge carriers without leading to excessive interface recombination or affecting Fermi-level splitting in the perovskite. The modification of indium-doped tin oxide, by adding an ultra-thin polyethylene-imine ethoxylated layer, produced surface dipoles and reduced its work function from 4.6 to 4.0eV. This change benefited electron transport from a TiO_2 electron-transport layer to the indium-doped tin oxide and the average power-conversion of the device was 16.6%. Among other additives to TiO_2, Zr^{4+} was the most effective and led to an average efficiency-increase from 15.0 to 15.8%, while a $Ti_{0.95}Nb_{0.05}O_x$ electron-transport layer helped to protect a perovskite from humidity.

Among the inorganic alternatives to a TiO_2 electron-transport layer are CdS, SnO_2, WO_3 and ZnO. The latter's electron mobility, of 205 to $300cm^2/Vs$, is better than that of TiO_2 but it tends to be chemically unstable. Zinc oxide has been used as an electron-transport layer to isolate the perovskite from an aluminium electrode in solar cells, leading to improved stability in air and to an average power-conversion efficiency of 14.6%. Tin oxide is also a promising electron-transport layer, with a wide band-gap and an electron mobility of $240cm^2/Vs$, and can be processed at low temperatures.

Among the most popular and efficient organic electron-transport materials which are used with perovskite solar cells are fullerene, and derivatives such as indene-C_{60}. Such materials are good for use as electron-transport layers in inverted planar perovskite solar cells because they can be processed at low temperatures, offer adjustable energy-level alignment and exhibit a good electron mobility.

Hole-Transport Layer

In the case of normally-structured perovskite solar cells, the hole-transport layer is commonly made from 2,2',7,7'-tetrakis[N,Ndi(4-methoxyphenyl)-amino]-9,9'-spirobifluorene (spiro-OMeTAD). The pure material unfortunately has a low conductivity, of the order of 10^{-6} to $10^{-5}cm^2/Vs$, and so lithium bis(trifluoromethanesulfonyl) imide (LiTFSI) and 4-tert-butylpyridine (TBP) are routinely added although TBP is corrosive and, complication on complication, LiTFSI is hygroscopic and so impairs the stability of the notoriously moisture-sensitive perovskites. Intercalation of Li^+ ions, migrating from the LiTFSI and into TiO_2 can also decrease the open-circuit voltage of solar cells. Moreover, cells which are spin-coated with spiro-OMeTAD tend to exhibit speedy degeneration of power-conversion efficiency under ambient conditions, and this has been attributed to the presence of a high number of pinhole defects in the coating; a sufficient number to create channels across the perovskite film. Effort is therefore expended on improving the stability of spiro-OMeTAD, and on avoiding charge-recombination at the perovskite|spiro-OMeTAD interface, by avoiding the creation of pinholes. Vacuum-evaporation is recommended for the preparation of pinhole-free spiro-OMeTAD films.

That material unfortunately suffers from the drawbacks of charge-carrier accumulation, degradation under ambient conditions and a high price. Alternative hole-transport layer materials have therefore been investigated, including CuI, Cu_2O, CuSCN, NiO_x, carbon and graphene nanotubes, methoxyphenylamine, polyfluorene and tetrathiafulvalene derivatives and poly(3-hexylthiophene-2,5-diyl),poly(triarylamine). Some of these are in fact superior to spiro-OMeTAD. For example, poly(3-thiophene acetic acid) (P3TAA) comprises –COOH groups which can interact with a perovskite via $–NH_3$ groups. Their diffusion across the P3TAA|perovskite interface can facilitate hole-injection from the perovskite to the P3TAA.

In the case of inverted planar perovskite solar cells, NiO_x and poly(3,4-ethylenedioxythiophene):polystyrene sulfonate (PEDOT:PSS) are most commonly used as the hole-transport layer. The latter is less satisfactory, due to poor electron-blocking, its acidity and its hydroscopic nature. The NiO_x-based choices meanwhile are associated with a smaller fill-factor (figure 6 and 7), due to low conductivity, but lead to a higher open-circuit voltage than does PEDOT:PSS. The performance of solar cells is described in terms of the 3 parameters: open-circuit voltage, V_{oc}, short-circuit current, I_{sc} and fill-factor. The maximum power-output of the device is given by their product.

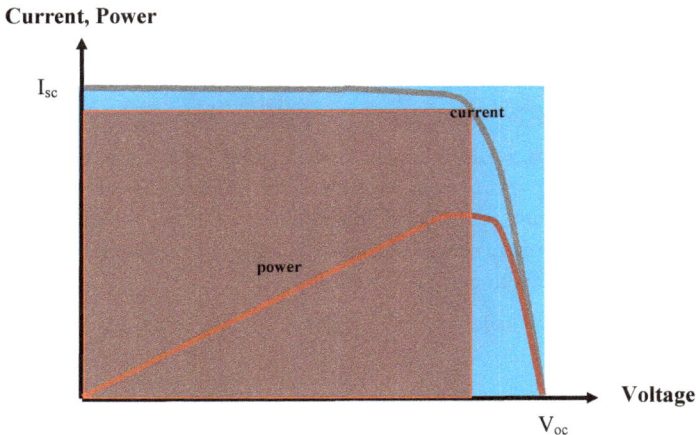

Figure 6. Pictorial definition of the fill-factor as the area of the largest rectangle which will fit within the I-V plot of a solar-cell as a fraction of the total area

Current, Power

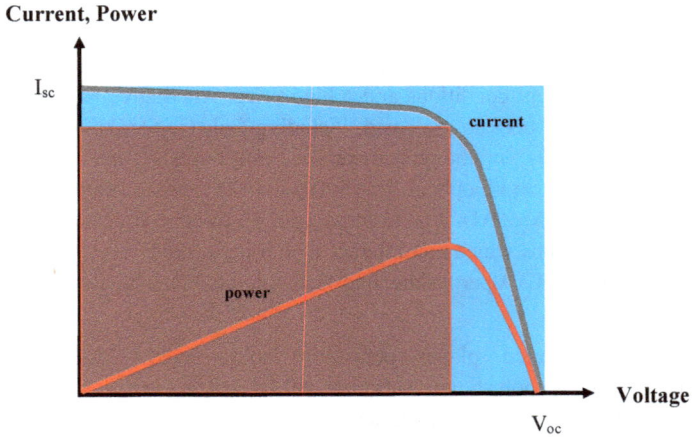

Figure 7. Current-voltage plot of a solar-cell as in the previous figure but having a smaller fill-factor

Heavy p-type doping of NiO can increase its electron conductivity and decrease the charge-transport resistance. As compared with solar-cells that involved PEDOT:PSS, the open-circuit voltage was about 100mV higher and the fill-factor had the extraordinary value of 0.827.

Interface Layers

The presence of interfacial layers in perovskite solar cells has several effects. Firstly, protection of the perovskite from moisture can improve long-term device stability. Secondly, slight resultant energy-level adjustments can reduce the energy-offset between adjacent layers and improve charge transport. Thirdly, the passivation of trap states in the perovskite can reduce photo-current hysteresis.

The layers are typically made from metals, forms of carbon and small organic molecules. Barium, calcium and LiF, with their low work-functions can be deposited prior to the evaporation of the aluminium, gold or silver electrode, and charge-carrier collection is thus enhanced by lowering the energy barrier. A LiF-buffer can improve the fill-factor and other cell characteristics. The deposition of MoO_x onto $CH_3NH_3PbI_3$ can raise the latter's valence-band to a higher binding-energy and thus allow better energy-alignment to a spiro-OMeTAD hole-transport layer. The oxides, Al_2O_3, Cr_2O_3, MgO, TiO_2, Y_2O_3

and ZnO, with their wide band-gaps are often used as barrier layers to decrease markedly the charge losses occurring across interfaces. They also prevent the corrosion of aluminium and silver electrodes which can result from the migration of halide ions. When gold is used as the top electrode, it can diffuse through a spiro-OMeTAD layer and into the perovskite even at temperatures as low as 70C. This can be prevented by depositing a chromium layer between the hole-transport layer and the gold electrode. Compounds such as Sb_2S_3 are also effective as barrier layers between TiO_2 and a perovskite. This sulfide can stabilize the solar cell against light-exposure, without requiring encapsulation, because any degradation apparently arises at the TiO_2|perovskite interface.

Perovskite solar cells often exhibit an anomalous hysteresis, of their current-density versus voltage plots, which has been attributed to various causes; including the action of mobile ionic species, or ferro-electricity. Such hysteresis is negligible in the case of certain cells which have a fullerene derivative as the electron-transport layer. It is found that such fullerene derivatives can passivate charge-traps on the surfaces and grain boundaries of the perovskite. For example, a spin-coated [6,6]-phenyl-C61-butyric acid methyl ester (PCBM) layer can cloak the perovskite, and permeate into it via the grain boundaries, thus decreasing surface charge recombination. The addition of just a small amount of PCBM to PbI_2 precursor solution could improve the quality of a perovskite by filling-in pinholes and vacancies between the perovskite grains.

The degree of hysteresis is particularly marked in n–i–p cells, possibly due to a depletion region and to pinning of the electron quasi-Fermi level across the TiO_2|perovskite interface. Fullerene derivatives can again modify the interface between a TiO_2 electron-transport layer and a perovskite in such a way as to reduce hysteresis, improve stability and increase electron transfer. A fullerene self-assembled monolayer can for instance improve electron transfer and passivate trap-states at the TiO_2|perovskite interface by anchoring TiO_2 surface groups via fullerene components at the perovskite surface. The C_{60} molecule is often chosen because of its high electron-affinity, while solvent-compatible cyano groups are good for pacifying oxygen vacancies and di-octyloxy chains.

Another carbon-based material which is used is graphene oxide. This has a 2-dimensional C–C structure, plus oxygen-containing groups such as carbonyl, carboxyl, lactone, phenol and quinone. Hydrophilic oxygen-containing groups reduce electron mobility and graphene oxide is therefore capable of suppressing electron recombination in perovskite solar cells. It can interact with a perovskite via a Pb–O bonding process which suppresses unsaturated lead-bonding at the interface. Two-dimensional C–C bonds in the graphene oxide can also absorb spiro-OMeTAD and thus increase the charge-collection due to an improved contact between the perovskite and the hole-transport layer.

Multi-walled and single-walled carbon nanotubes have also been used to collect holes from perovskite solar cells. A 5nm single-walled carbon nanotube film at the interface between a perovskite and spiro-OMeTAD can ensure sub-ps hole-extraction from a perovskite.

Self-assembled monolayers can modify the interface between the electron-transport layer and the perovskite, and substrates which are coated with amino or ammonium groups can encourage the growth of smooth crystalline perovskite films. This behaviour is attributed to the effect of hydrogen bonding or of electrostatic interactions between the functional groups and the perovskite structure. Self-assembly of amino-silane monolayers can occur via the formation of covalent bonds by condensation-reaction between alkoxy groups in silane and hydroxyl groups at a TiO_2 surface. Thiols can also be used in self-assembled monolayers, at a TiO_2|perovskite interface, so as to improve the performance and stability of perovskite solar cells. Self-assembled monolayers can also render a perovskite cell more resistant to moisture.

The thermal evaporation of bathocuproine, C_{60}, calcium or LiF onto a PCBM layer is often used to improve the fill-factor to a great extent, but this requires high-vacuum processing. It has been shown however that organic molecules can serve as buffer layers between PCBM and a metal electrode. Those materials can be applied to the PCBM, by means of low-temperature solution-processing, and lead to a reduced leakage current and a high fill-factor.

Finally, quantum-dots have found some applications in the field of lead-halide perovskite solar cells. When graphene quantum-dots were inserted as an ultra-thin layer between a TiO_2 electron-transport layer and a perovskite, the effectiveness of the resultant solar cell was increased due to rapid electron-extraction and the power-conversion efficiency was increased from 8.81 to 10.15%. Ultra-fast transient absorption spectroscopy revealed that shorter electron-extraction times obtained across the perovskite|graphene-quantum-dot|TiO_2 interface (90 to 106ps) and across the perovskite|TiO_2 interface (260 to 307ps) thus suggesting the occurrence of accelerated electron injection caused by the quantum-dots.

Preparation

The mechanism involved in the conversion of a precursor ink into a perovskite film using antisolvent-induced crystallization has been studied[30] using *in situ* X-ray diffraction during blade-coating and antisolvent deposition of methylammonium lead iodide perovskite. It was found to be essential to add the antisolvent before any formation of an intermediate crystalline phase. Instead of heterogeneous nucleation and conversion via a crystalline intermediate phase, antisolvent-assisted annealing leads to rapid direct crystallization and highly regular smooth films.

Vacuum-deposited inorganic perovskite thin films can be prepared[31] by the co-sublimation of cesium halide and lead halide precursors under a high vacuum. The hygroscopic nature of these halides has to be taken into account in order to control accurately the stoichiometric ratio in the finished films. By integrating vacuum-sublimed, hole-transport and electron-transport layers, entirely vacuum-deposited solar cells can be made, and devices made using stoichiometrically-balanced precursors lead to a much better performance.

A method was proposed[32] for the preparation of the perovskites via low-pressure vapour-assisted solution. This began with the synthesis of CH_3NH_3I and CH_3NH_3Br from methylamine and HI or HBr. Pinhole-free continuous $CH_3NH_3PbX_3$ (X = I, Br and/or Cl) films were then made by using the low-pressure vapour-assisted solution process. This involved spin-coating an homogenous layer of lead halide precursor onto a substrate and conversion of that layer into $CH_3NH_3PbI_{3-x}Br_x$ by exposing the substrate to the vapour of a mixture of CH_3NH_3I and CH_3NH_3Br under reduced pressure at 120C. The slow diffusion of methylammonium halide vapour into the lead halide precursor then led to the slow controlled growth of a continuous pinhole-free perovskite film. This process permitted the preparation of compositions of the form, $CH_3NH_3PbI_{3-x}Br_x$ (x = 0 to 3). The band-gap could thereby be varied from 1.6 to 2.3eV.

Room-temperature preparation of pinhole-free perovskite films of high crystallinity is possible via the use of nanostructured PbI_2L_x intermediates, where L is a ligand. By using this method, solar cells with a power-conversion efficiency of 17.21% could be created[33]. Perovskites which were prepared at room temperature had a power-conversion efficiency of less than 16%.

Various concepts for the preparation of methylammonium lead halide perovskite particles were considered[34]. One was based upon the wet-chemistry preparation of the particles by adding perovskite precursor solution to an anti-solvent so as to provoke the precipitation of perovskite particles in the solution. Another was based upon the milling of a blend of perovskite precursors in dry form so as to allow conversion of the precursors to perovskite particles. The final idea was based upon the atomization of perovskite solution by using a spray nozzle and the introduction of the spray droplets into a hot-wall reactor. This was found to be the most successful method for the preparation of impurity-free perovskite particles and the deposition of perovskite thin films.

As well as the preparation of the perovskites themselves, attention also has to be paid to the preparation of the other components of the solar cells. A colloidal method was described[35] for TiO_2 preparation and the formation of compact and conformal TiO_2 layers on transparent conducting substrates. The colloid which was thereby created had a particle size of about 5nm and could remain stable for more than a year. The method was used to create hybrid perovskite solar cells in ambient air with a relative humidity of 40% at room temperature. The important hole-blocking ability of a one-step TiO_2 colloid-based compact layer was confirmed by a resultant power-conversion efficiency of 8.08%, observed for active areas of only $0.07cm^2$ and of 7.52% for active areas of $0.23cm^2$. When using 2-step sequential deposition, a power-conversion efficiency of 4.61% was observed for an active area of $0.64cm^2$. Magnesium-doped TiO_2 nanorods were prepared[36] from colloidal titanate by means of microwave hydrothermal reaction. The use of the nanorods, with a high conduction-band edge, as an electron-extracting material for ammonium lead halide perovskite solar cells increased the open-circuit voltage by up to 215mV.

Cesium-Doped Lead-Halide Perovskites

It is suggested that the cesium-doping of organic-inorganic metal-halide perovskites may be the best way to stabilize solar cells. A semiconductor, with a band-gap of the order of 1.8eV, which is capable of large-scale processing could permit the manufacture of highly-efficient cells at a cost of a few dollars per square metre. These solution-processable organic-inorganic halide perovskites can serve as absorbers in single-junction solar cells and it is possible to vary the band-gap of $(CH_3NH_3)Pb(Br_xI_{1-x})_3$, for example, from 1.6 to 2.3eV by controlling the halide concentration. On the other hand, optical instability arising from photo-induced phase-segregation limits the voltages which can be expected from those compositions having suitable band-gaps for tandem applications. It has even been questioned[37] whether organic cations are essential to the superiority of halide perovskites. These materials also unfortunately undergo thermal degradation at temperatures which are likely to be encountered during processing and operation. By replacing volatile methylammonium with cesium, it is possible to produce mixed-halide absorbers having a stabilized photoconversion efficiency of perhaps 6.5% and a band-gap of 1.9eV. A systematic study[38], under oxygen- and moisture- free conditions, of the thermal and photochemical degradation of materials of the form, $APbX_3$, with X being iodine or bromine, A being CH_3NH_3 and cesium cations, showed that the most common hybrid material, $CH_3NH_3PbI_3$, is intrinsically unstable with respect to heat- and light-stressing and is thus unsuitable for practical solar-cell use. Cesium-based all-inorganic lead halides were far superior but, although they enjoy a relatively high phase-stability they can still suffer from a low photo-electric conversion efficiency.

Understanding the roles played by the organic molecule and the inorganic lattice is essential when attempting to optimize the physical properties, especially with regard to the recombination mechanisms. Molecular dynamics and frozen-phonon methods have revealed sub-ps anharmonic fluctuations in the entirely inorganic perovskite, $CsPbI_3$. These fluctuations, combined with spin-orbit coupling, have an effect upon the electronic band structure, as monitored[39] via a dynamic Rashba effect. That is, the conduction-band has 2 parts while splitting of the valence-band is negligible. Thus the lowest-energy optical transition is indirect while direct transitions at higher energies remain possible. Under certain conditions, spatial disorder can quench the Rashba effect. In the case of temporal disorder, a dynamic Rashba effect is detected which is similar to that found in $CH_3NH_3PbI_3$, and which is still appreciable in spite of thermal disorder and the absence of organic cation motion. The spin texture which was associated with the Rashba splitting could not be responsible for a consistent reduction in the recombination rates, even though the spin mismatch-between the valence and conduction bands increases with the ferroelectric distortion causing Rashba splitting. The hybrid lead halide perovskites are essentially semiconductors which exhibit very large optical absorption coefficients in the visible range, long diffusion-length photo-excited charge-carriers and lengthy excited-state lifetimes, but a persistent question is whether the optical band-gap is direct or indirect. It has been suggested that Rashba spin–orbit coupling can give rise to an indirect gap which is some tens of meV lower in energy than is the direct one which is usually assumed to exist. Radiative recombination rates in hybrid perovskites have been measured[40], as a function of temperature, by measuring the instantaneous photoluminescence intensity under pulsed excitation. The results showed that radiative recombination became more rapid with decreasing temperature, as in the case of all direct-bandgap materials; the opposite of that which was expected for 3-dimensional Rashba semiconductors.

The prevention of halide-ion exchange between $CsPbBr_3$ and $CsPbI_3$ nanocrystals, by means of capping with $PbSO_4$-oleate, permits the deposition of different perovskite nanocrystals as aligned arrays on surfaces without any intermixing[41]. The electrophoretic deposition of such capped $CsPbX_3$ nanocrystals, suspended in hexane solution, onto mesoscopic TiO_2 films has permitted the use of device architectures involving single or multiple layers of perovskite film. There is first a linearly organized arrangement of nanocrystals in hexane, followed by the deposition of larger rods which are about 500nm in length. Although not strictly relevant to solar-cell use, this makes it possible to construct films having a tunable luminescence that includes white. It is interesting to note that nanocrystals of the essentially non-fluorescent zero-dimensional cesium lead halide perovskite, Cs_4PbBr_6, exhibit distinct emissions under a pressure of 3.01GPa, and the intensity increases markedly under further compression[42]. This has been attributed to enhanced optical activity and to an increase in the binding-energy of self-trapped excitons under compression. The effect can be further traced to the large distortion of $[PbBr_6]_4$ octahedra which occurs during structural phase-changes. Compression can thus be useful in increasing the photoluminescence efficiency.

The high refractive index of lead halide perovskites causes their interfaces with air to have a reflectivity which is greater than 15%. An experimental comparison has been made[43] of transient transmissions from lead halide perovskite films and weakly quantum-confined nanocrystals of cesium lead iodide perovskite, showing that changes in absorption rather than in reflection determine the results of transmission measurements of thin films of these materials. That is, none of the observed characteristic spectral signatures are entirely due to, or greatly affected by, changes in the sample's reflectivity.

The hole-extraction ability of 1-ampinopyrene from $CsPbBr_3$ and $CsPbI_3$, with their different band-gaps, has been investigated[44]. An efficient quenching of the photoluminescence of nanocrystalline samples, with little change in the photoluminescence time-profile in the presence of 1-ampinopyrene, indicated a static interaction of the interacting species. Band-alignment of the nanocrystals, relative to the highest occupied and lowest unoccupied molecular-orbital energy-levels of 1-ampinopyrene, revealed rapid hole-transfer from the photo-excited nanocrystals to 1-ampinopyrene. The hole-transfer time-constants were estimated to be 120 and 170ps for $CsPbBr_3$ and $CsPbI_3$ nanocrystals, respectively. The rapid hole extraction appeared to result from strong anchoring of 1-ampinopyrene to the nanocrystal surface and easy transfer of the hole through the conductive pyrene framework.

A reproducible low-cost method[45] for synthesizing high-quality cesium lead halide perovskite nanocrystals involves the direct heating of precursors in octadecene, under air. The particle-size and composition of the as-prepared nanocrystals can be adjusted simply by varying the reaction temperature. The emission-peak of the nanocrystals can also be easily varied from the ultra-violet to the near-infrared (360 to 700nm) range, and the quantum-yield of the product can attain 87%.

A simple room-temperature (25C) ligand-mediated re-precipitation method can be used[46] to control the shape of colloidal $CsPbX_3$ nanocrystals so as to yield spherical quantum-dots, nanocubes, nanorods or nanoplatelets. The spherical quantum-dots led to photoluminescence quantum-yields of up to 80%, and the corresponding emission-peak positions ranged from 380 to 693nm. The other shapes were produced using various organic acids and amine ligands in the re-precipitation process. The shape-dependent photoluminescence decay lifetimes ranged from several hundred to tens to hundreds of nanoseconds. The all-inorganic perovskites are plagued by low-temperature phase instability and by degradation in the presence of moisture and polar solvents. When perovskite quantum-dots with a size of about 10nm were prepared[47] using organic ligand assisted means, variation of the composition of $CsPbBr_xI_{3-x}$ suggested that $CsPbBr_{1.5}I_{1.5}$ could be an alternative quantum-dot material. The photoluminescence decay had a comparable average lifetime, and photovoltaic devices made from $CsPbBr_{1.5}I_{1.5}$ exhibited a power-conversion efficiency of about 7.94%, with an open-circuit voltage of about 1.0V, a fill-factor of 0.70 and negligible hysteresis. The halide content markedly affects non-radiative electron-hole recombination in all-inorganic perovskite quantum-dots. Replacement of half of the bromine ions with iodine ions, in a $CsPbBr_3$ quantum-dot, lengthens the charge-carrier lifetime by a factor of 5[48]. Complete replacement lengthens the lifetime by a factor of 8. Doping with iodine decreases the non-adiabatic charge-phonon coupling because the iodine ions are heavier and slower than bromine ions, and because the overlap between the electron and hole wave-functions is reduced. Non-radiative electron-hole recombination tends to occur at a ns-rate, due to a non-adiabatic coupling of less than 1meV and a coherence-time of less than 10fs.

The possibility of modifying metal halide perovskite quantum-dot surfaces is essential for their practical application, but is difficult because conventional techniques lead to transformation or dissolution of the crystal. Ligand-exchange methods can electronically couple the quantum-dots while retaining a nanocrystalline size and stabilizing the corner-sharing structure of the PbI_6^{4-} octahedra. By individually targeting the anionic oleate and cationic oleylammonium ligands, it is found[49] that ambient moisture helps by hydrolysing methyl acetate to acetic acid plus methanol. The acetic acid replaces native oleate ligands to yield quantum-dot surface-bound acetate and free oleic acid. The native oleylammonium ligands persist during film deposition and are then exchanged during final treatment with formamidinium. The final treatment leads to more strongly coupled quantum-dots and eventual transformation into bulk metal halide perovskite film.

Monodisperse colloidal nanocubes of fully inorganic cesium lead halide perovskites, having an edge-length of 4 to 15nm, can be made[50] by using inexpensive commercial precursors. By means of compositional variations, the band-gap energies and emission spectra can be easily varied over the visible spectral range of 410 to 700nm. The photoluminescence of $CsPbX_3$ nanocrystals is characterized by narrow emission line-widths of 12 to 42nm, a wide range of colours, quantum yields of up to 90% and radiative lifetimes ranging from 1 to 29ns. It has been shown experimentally[51] that a carrier-density dependence of the band-gap renormalization, and an effective hot-phonon bottleneck, occur in $CsPb(Cl_{0.20}Br_{0.80})_3$ mixed-halide nanocrystals. The optical response changed markedly, over a spectral range of hundreds of meV at high carrier densities, due to appreciable band-gap renormalization. The band-gap renormalization constant of $CsPb(Cl_{0.20}Br_{0.80})_3$ nanocrystals was calculated to be about 6.0×10^{-8} eVcm. An efficient hot-phonon bottleneck was detected at a carrier-density of 3.1×10^{17}/cm^3, and this slowed thermalization by an order-of-magnitude.

Cesium lead halides of the form, $CsPbX_3$ where X is iodine or bromine, can be prepared using a simple cold-sintering technique[52] to yield single-phase samples with a Pnma structure. Negligible changes in the crystal structure and band-gap have been noted up to 2400h later; thus indicating adequate stability under ambient conditions. The band-gap was comparable to that of materials prepared using other means.

Dip-coating is a very efficient solution-deposition technique for preparing perovskite quantum-dots. In this process, the withdrawal speed is the key factor because it determines the film thickness and possibly its quality. The optimum withdrawal speed for producing optically smooth $CsPbX_3$ films on a GaAs substrate was found[53] to be 10mm/s.

Spin-coating can be used[54] to deposit a thin layer of dodecylamine hydro-iodide onto the surface of $CH_3NH_3PbI_xCl_{3-x}$ perovskite and change the surface from being hydrophilic to being hydrophobic. A markedly improved fluorescence intensity and a longer fluorescence lifetime are exhibited by the modified films, and this is attributed to a sharp reduction in the number of structural defects. Compatibility between the perovskite and the hole-transfer layer is also improved, leading to more efficient hole-collection from the perovskite layer. Solar-cells which are made from the modified perovskite films enjoy a much better resistance to humidity stability and a superior photo-electron conversion efficiency.

In connection with energy-storage applications, a $CsPbBr_{2.9}I_{0.1}$ perovskite solar cell was integrated into an asymmetrical super-capacitor device[55]. The cell retained 70% of its efficiency after a week of storage in a dark humidity-controlled environment, and 33% efficiency under ultra-violet light and 24h of exposure to air with a relative humidity of more than 80%.

Cesium lead halide thin films can be prepared in 'mille-feuille' form by rapid alternating deposition[56]. When integrated into solar cells, they can exhibit a power-conversion efficiency of 13.0%. The use of these devices for environmental-light energy-harvesting would give a power-conversion efficiency of 33.9% under 1000lux fluorescent-light illumination.

The effect of cesium incorporation upon the minority-carrier recombination lifetime of cesium-methylammonium lead halide perovskite, $Cs_x(CH_3NH_3)_{1-x}PbI_{3-x}Br_x$, thin films is such that the lifetime of the as-deposited film decreases with increasing cesium concentration[57]. Mixed-cation perovskite films are more stable, with lifetimes of 15 to

20μs following 9h of ambient exposure as compared with 6 to 13μs just after deposition. Exposure to methylamine vapour improves the morphology of as-deposited films, with their becoming more oriented along the (110) direction. They are then even more stable under ambient conditions, with $Cs_{0.10}(CH_3NH_2)_{0.90}PbI_{2.90}Br_{0.10}$ films having lifetimes of almost 50μs following 9h of ambient exposure.

A single-step chemical vapor deposition method can be used[58] to grow cesium lead halide microcrystals. These have a square-platelet morphology and are highly crystalline, with $CsPbCl_3$, $CsPbBr_3$ and $CsPbI_3$ having tetragonal, monoclinic and orthorhombic structures, respectively. Platelets are well-faceted, and are of the order of 10 to 50μm in size, with a thickness of about 1μm. Their very smooth surfaces suggest an absence of grain boundaries and a monocrystalline nature. Their optical emissions are uniform and intense, consistent with the expected band-edge transition, and the photoluminescence is greater around the platelet edges, indicating a wave-guiding effect. The well-defined geometry and ultra-smooth surfaces mean that square platelets can exhibit whispering-gallery phenomena with a quality-factor of up to 2.863. The micro-platelets can be easily grown on substrates such as silicon, graphene and molybdenum disulfide.

Femtosecond transient absorption spectroscopy of the interfacial charge-transfer of $CsPbX_3$ films with various halogen-ratios, in contact with TiO_2, was used[59] to study post-photoexcitation hot-carrier cooling, free-exciton formation, electron transfer and charge recombination. The time-constant of interfacial electron transfer depended upon the location of the trap-state of the perovskite and the relative energies of the conduction bands in the perovskite and TiO_2, and the time-constant of charge recombination depended upon electron-hole interactions. An increased iodine content in $CsPbX_3/TiO_2$ systems increased the time-constants of electron-transfer and charge recombination; implying that all-inorganic $CsPbX_3$ perovskites having a high iodine content were likely to improve the power-conversion efficiency of solar cells. Fluorine can also be introduced so as to control the bulk heterostructures of $CsPbBrI_{2-x}F_x$ compounds and impart considerably improved power-conversion efficiencies and phase stability. An α-phase/δ-phase heterojunction encourages the efficient dissociation of excitons and of a charge transport driven by matched energy-band offsets. The lifetime of the charge carriers is lengthened because of retarded charge-recombination, as revealed by time-resolved photoluminescence data. It leads to an improved short-circuit current-density. Because of the Goldschmidt geometrical tolerance factor, the partial replacement of the iodine ions by smaller fluorine ions stabilizes the α-$CsPbBrI_2$ structure. The composition, $CsPbBrI_{1.78}F_{0.22}$, in particular[60], with an α-phase/δ-phase heterostructure exhibits a power-conversion efficiency of up to 10.26% combined with an appreciable structural stability with respect to moisture and time.

Doping of $CsPbIBr_2$ with manganese also improves[61] the properties of the all-inorganic perovskite. As the Mn^{2+} ion concentration in materials of the form, $CsPb_{1-x}Mn_xI_{1+2x}Br_{2-2x}$, is increased, the band-gap decreases from 1.89 to 1.75eV. At some manganese concentrations, the crystallinity and morphology are superior to those of the undoped material. These advantages off-set an energy-loss in hole-transfer and facilitate charge-transfer, so that solar-cells which are based upon these films give a better photovoltaic performance than do undoped $CsPbIBr_2$ perovskite films. The latter composition attains a power-conversion efficiency of 6.14%, while $CsPb_{1-x}Mn_xI_{1+2x}Br_{2-2x}$ cells give a power-conversion efficiency of 7.36% when x is equal to 0.005; a 19.9% improvement in power-conversion efficiency. Encapsulated $CsPb_{0.995}Mn_{0.005}I_{1-01}Br_{1.99}$ cells also exhibit a good stability in ambient atmospheres, with the power-conversion efficiency falling by only 8% after more than 300h. The most recent work has shown[62] that by modifying the interface between the hole-transport layer and the perovskite light-absorption layer, and by optimizing the former layer with regard to improving energy alignment, it is possible to control growth, reduce carrier recombination, promote carrier injection/transport and improve both the cell's power conversion efficiency and its moisture stability. When testing using a positive bias scan, a record-breaking improvement in power conversion efficiency (9.49%) for such cells was obtained. Other measurements showed that passivated $CsPbIBr_2$-based inverted solar cells could maintain 86% of their initial efficiency following 1000h in ambient air of 65% relative humidity.

Time-domain density functional theory and non-adiabatic molecular dynamics methods have indicated[63] that a photo-induced localized polaron-like hole in $CsPbBr_3$ can greatly delay non-radiative electron-hole recombination, as compared with structures having delocalized free charges. This is attributed to the fact that localized charge-carriers decrease the overlap between electron and hole wave functions and decrease non-adiabatic coupling by a factor of 6. Polaron formation also increases the band-gap of $CsPbBr_3$, further slowing recombination. A smaller non-adiabatic coupling and larger band-gap combine, with a longer decoherence time, to extend the recombination time to some tens of nanoseconds.

A full-coverage all-inorganic $CsPbBr_3$ perovskite film can be produced[64] by introducing the small organic molecule, 1,3,5-tri(m-pyrid-3-yl-phenyl)benzene (TmPyPB), as an additive to solutions. A device which was made by using a $CsPbBr_3$:TmPyPB film as a light-emitting layer exhibited a maximum brightness of $22309cd/m^2$, a maximum current efficiency of 8.77cd/A and an external quantum-efficiency of 2.27%. These figures were 8.6, 10.2 and 10.3 times those of a plain $CsPbBr_3$-based light-emitting diode, respectively. The improved electroluminescent was attributed to a lower current-leakage, due to the full coverage, and to increased electron-transport in the $CsPbBr_3$:TmPyPB perovskite film.

In perovskite-based dye-sensitized materials, the $CsPbI_3$ perovskite is usually paired with a hole-transport material, such as spiro-OMeTAD. This extracts a hole from the photo-excited perovskite and generates free charge-carriers. Two competing charge-transfer pathways can be identified at the interface between a perovskite and spiro-OMeTAD. One route is termed, through-bond, while the other is termed through-space. The former route involves a segment of spiro-OMeTAD which contains methoxy linking groups. The through-space case involves a segment of spiro-OMeTAD with linking groups removed. Four atomistic models have been considered[65]. In one of them, a periodic cesium lead iodide perovskite nanowire was paired with the through-space route. In another, a periodic perovskite nanowire was paired with the through-bond route, with the linking groups forming a coordination bond with the surface of the nanowires. In a third model, a perovskite thin film was paired with the through-space route and, in the fourth model, a perovskite thin film was paired with the through-bond route. The charge-transfer dynamics of the rates of electron/hole relaxation and relaxation paths, were calculated by using a Redfield reduced density matrix formalism. The terminal surface, lead-iodine or cesium-iodine, of the perovskite was significant with regard to energetic alignment at the perovskite/dye interface, due to band-bending. The upper and lower bounds on experimental results were shown to correspond to the through-bond through-space pathways, respectively.

The addition of cesium ions to $CH_3NH_3PbI_3$ significantly improves its thermal stability, and a non-encapsulated perovskite device could retain[66] an energy-conversion efficiency equal to about 75% of its original value in the case of $(CH_3NH_3)_{0.85}Cs_{0.05}PbI_3$ after annealing for 80min at 140C in an atmosphere with a relative humidity of less than 30%.

The stability with regard to humidity can be greatly improved by separating quantum dots using a SiO_2 shell, but a poor thermal stability of $CsPbX_3$ quantum-dots continues to be associated with the intrinsically low formation energies of perovskite lattices. An effective solution is to use Mn^{2+} additions to stabilize the perovskite lattice of $CsPbX_3$ quantum-dots at up to 200C under ambient atmospheric conditions. The significantly improved thermal stability and optical performance of $CsPbX_3$:Mn^{2+} quantum-dots[67] can be attributed to an increased formation energy.

The exciton binding energy and reduced mass of $CsPbX_3$ (X = I and/or Br) perovskites have been determined[68] by means of magneto-transmission measurements. A continuous temperature variation of the band-gap between 4 and 270K suggested that fully inorganic perovskites, unlike hybrids, do not undergo structural phase transitions. At low

Materials Research Forum LLC

https://doi.org/10.21741/9781644900819

temperatures, where motion of the organic cation is frozen, the dielectric screening mechanism was essentially the same for both hybrid and inorganic perovskites and was dominated by the relative motion of atoms within the lead halide cage.

Nanowires of $CsPbI_3$, with a diameter of 50 to 100nm, were deposited[69] onto fluorine-doped tin oxide glass by using a simple solution-dipping process. They were then transformed into $CsPbBr_3$ nanowires via solution-phase halide-exchange. A phase-change from a non-perovskite to the perovskite structure occurred during the replacement of iodine ions by bromine ions. The as-formed $CsPbI_3$ and $CsPbBr_3$ nanowires were incorporated into perovskite solar cells, leading to power-conversion efficiencies of 0.11 and 1.21%, respectively. The $CsPbBr_3$ nanowire cell retained 99% of its initial power-conversion efficiency after 5500h of aging.

Cesium lead halide perovskite solar cells exhibit an improved performance and stability after incorporating potassium cations. A cell which was based upon $Cs_{0.925}K_{0.075}PbI_2Br$ attained a power-conversion efficiency of 10.0%, with a lengthened operational lifetime in air[70].

A systematic investigation has been made[71] of the geometric structures of high-angle tilt grain boundaries in perovskites of the form, $CsPbX_3$, where X is bromine, chlorine or iodine. This was based upon the coincidence site lattice model, and refined by means of lattice shifts and expansions. The boundaries did not generate mid-gap states, due to the large distance between unsaturated atoms and atomic reconstructions in the grain-boundary region. The grain boundaries could instead introduce other, very shallow, states near to the valence-band edge, and those could hinder hole-diffusion.

A study was made[72] of the effect of iodine substitution on the structural and optical properties of solution-processed $CsPbBr_{3-x}I_x$ (x = 0 to 1) thin films. A Pawley fit indicated that the mixed perovskite crystallized in the orthorhombic Pnma space group. X-ray diffraction results indicated a shift to lower angles as the fraction of iodine increased; thus indicating an expansion of the lattice. The films exhibited a very high absorbance in the visible and short infra-red range. The band-gap varied from 2.38 for $CsPbBr_3$ to 2.17eV for $CsPbBr_2I$, in accord with Vegard's law. The $CsPbBr_3$ films emitted a bright photoluminescence, with a maximum at 530nm. The $CsPbBr_{3-x}I_x$ films retained an excellent stability after aging for 48h at a relative humidity of about 60% at relatively high temperatures.

It has been commonly remarked that many quite simple processing methods produce surprisingly reliable batches of metal halide perovskite thin films. The reason has now been tentatively traced by studying[73] the crystal and electronic structures of $CsPbBr_3$ samples having various cesium contents. At sub-stoichiometric concentrations of the latter, there occurred large variations in the electronic structure due to quite small variations in the cesium content. A critical point was detected however beyond which large variations in the chemical composition led to small associated changes in the valence and conduction-band energies. This revealed a remarkable insensitivity to large changes in stoichiometry which was very unlike that exhibited by certain compound semiconductors, for example.

Although it is possible to vary the band-gap of $CH_3NH_3Pb(Br_xI_{1-x})_3$ from 2.3 to 1.6eV by adjusting the halide concentration, an optical instability due to photo-induced phase-segregation limits the voltage that can be obtained from those compositions having suitable band-gaps for tandem applications. By replacing the volatile methylammonium cation with cesium[74], it is possible to produce a mixed halide absorber which exhibits an improved optical and thermal stability; with a stabilized photo-conversion efficiency of 6.5% and a band-gap of 1.9eV.

Green-emitting devices have been created[75] by using inorganic cesium lead halide perovskite nanocrystal emitters. The introduction of a thin film of perfluorinated ionomer between the hole-transport layer and the perovskite emissive layer markedly increased the hole-injection efficiency. The perfluorinated ionomer layer also prevented charging of the perovskite nanocrystal emitters and thereby preserved their emissive properties, leading to a three-fold increase in the peak brightness to $1377cd/m^2$. The full-width at half-maximum of the symmetrical emission peak (colour coordinates: 0.09 and 0.76) was 18nm.

It has recently been suggested[76] that the same photo-physical properties of perovskites which have already proved useful for photovoltaic applications, should also be of use in the field of photo-redox organic synthesis, given that both applications rely upon charge-separation and charge-transfer. It was shown that perovskite nanocrystals were very promising as photocatalysts for promoting organic reactions such as C–C, C–N and C–O bond-formation. The stability of $CsPbBr_3$ in organic solvents, and the ease with which its band-edge can be varied, gives the perovskite a wide scope of action.

The utility of opto-electronic materials can be greatly impaired by the existence of efficient pathways for non-radiative recombination. The lead halide perovskites are popular because they are simple to synthesize, easily absorb visible light and have a low incidence of non-radiative recombination. A theoretical study[77] of possible pathways for non-radiative recombination in a generic lead halide perovskite, $CsPbBr_3$, identified a set of conical intersections in a molecule-scale cluster model, Cs_4PbBr_6, and a nanoparticle-scale model, $Cs_{12}Pb_4Br_{20}$, of the $CsPbBr_3$ surface. The energies of the minimal-energy conical intersections, as corrected for dynamic electron correlations and spin-orbit coupling, were far above the bulk band-gap of $CsPbBr_3$. This suggested that those intersections did not provide efficient pathways for non-radiative recombination. Analysis of the electronic structure at the intersections suggested that the ionic nature of the bonds in $CsPbBr_3$ played a role in setting the high energies of the conical intersections. All of the lowest-energy intersections involved charge-transfer over long distances; either across a dissociated bond or between neighbouring unit cells.

A comparison was made[78] of thin films of the metal halide perovskite, $APbX_3$, of bulk $CH_3NH_3PbI_3$ prepared by spin-coating of the precursors in solution and of $CsPbBr_3$ colloidal nanoparticles. The $CH_3NH_3PbI_3$ thin films, grown using a single-step method, comprised separate grains having random orientations. The growth method led to the formation of tetragonal perovskite thin films which adhered well to an underlying TiO_2 layer and had a photoluminescence emission band which was centered on 775nm. Perovskite thin films were based upon $CsPbBr_3$ colloidal nanoparticles that were preserved by the deposition process, although small gaps existed between adjacent nanoparticles. The crystal structure of the $CsPbBr_3$ nanoparticles was cubic, and this benefited the optical properties due to an optimum band-gap. The absorption and photoluminescence spectra of the thin films and of the colloidal solution of $CsPbBr_3$ nanoparticles were very similar. This reflected good homogeneity of the thin films, and an absence of nanoparticle-aggregation. The material unfortunately exhibited a tendency to decompose due to lead segregation.

First-principles calculations predict that the incorporation of rubidium and potassium in suitable ratios will greatly stabilize $CsPbI_2Br$ perovskites. It was shown[79] that 2-dimensional 1nm-thick orthorhombic $CsPbI_3$ exhibits phenomenal stability.

The perovskites have recently been shown[80] to exhibit an unusual temperature dependence of the charge-carrier lifetime. In the particular case of $CsPbBr_3$, the unusual behaviour was traced to the highly anharmonic nature of atomic motion in the perovskite. As the temperature is increased, the perovskite structure deforms appreciably, resulting in a tilting of the octahedral units, and undergoes large-scale anharmonic departures from the equilibrium geometry. The electronic energy-gap thus increases, and phonon-induced coherence-loss within the electronic sub-system accelerates. These effects slow the non-radiative electron-hole recombination which is the main limitation on the efficiency of perovskite-based devices. This increase in charge-carrier lifetime with temperature is clearly beneficial in practice because solar-cells naturally tend to heat up during use. The behavior of all-inorganic halide perovskites differs from that of hybrid organic-inorganic perovskites: the latter exhibit a further disorder which is associated with reorientation of the asymmetrical organic cations.

Materials Research Forum LLC
https://doi.org/10.21741/9781644900819

A comparison was made[81] of the electron-irradiation responses of all-inorganic $CsPbI_3$ and organic-inorganic $CH_3NH_3PbI_3$ perovskite thin films. Under continuous irradiation at high dose-rates, the $CsPbI_3$ exhibited quite rapid morphological changes and a compositional degradation which was attributed to radiolysis, plus perhaps beam-heating and Coulomb forces. In free-standing polycrystalline $CH_3NH_3PbI_3$ thin films, there was appreciable structural damage, mainly near to the periphery of the irradiated region, while other areas remained largely intact. An electron beam-induced electric-field effect was thought to be the main reason for the unusual damage distribution. The apparent build-up of induced electrical fields in $CH_3NH_3PbI_3$ was thought to be related to its unusual structural properties, such as the rotational dynamics of the organic cations.

The ABX_3 perovskites are usually prepared in an inert atmosphere, as their properties are greatly affected by the humidity and temperature of the environment. Bulk samples of $CH_3NH_3PbI_3$ and $CsPbI_3$ (table 5) have been processed[82] at ambient temperature and pressure in order to determine their thermodynamic parameters, and to clarify the role played by the methylammonium ion in lead halide perovskites. The thermodynamic parameters suggested that $CsPbI_3$ is more heat-sensitive than is $CH_3NH_3PbI_3$, due to a low specific heat. A greater degree of bending of the Pb-I bond, as compared with stretching of the Pb-I bond in $CH_3NH_3PbI_3$ and the occurrence of only stretching modes in $CsPbI_3$ along the c-axis, increased tetragonal strain in the latter material. The concurrence of bending of the Pb–I bond, and torsion of the CH_3NH_3–I bond, leads to a decrease in the band-gap in $CH_3NH_3PbI_3$ as compared with $CsPbI_3$. This also increases the possibility of electrons and holes in the conduction and valence bands, respectively. A more strained structure led to the decrease in the specific heat of $CsPbI_3$.

Table 5. Structural parameters of $CH_3NH_3PbI_3$ and $CsPbI_3$

Parameter	$CH_3NH_3PbI_3$	$CsPbI_3$
structure	tetragonal	orthorhombic
space group	I4/mcm	Pnma
a	8.8687Å	10.4716Å
b	8.8687Å	4.8049Å
c	12.6706Å	17.7983Å
c/a	1.43	1.7
cell-volume	996Å^3	895Å^3

The gain coefficients of colloidal $CsPbBr_3$ nanocrystals were measured[83] by means of broadband transient absorption and ultra-fast fluorescence spectroscopy. The optical gain in such nanocrystals was supported by stimulated emission from free carriers, and not from excitons or bi-excitons. Even in the case of the fully inorganic lead halide perovskite, a cooling bottleneck impaired the development of nett stimulated emission at high excitation densities.

All-inorganic lead halide perovskite $CsPbBr_3$ thin films, with micron-sized grains, which were prepared by heat-spraying using a $CsPbBr_3$-saturated solution had large areas, a low defect content and high stability[84]. The grain size of the films ranged from 1 to 5μm, and these micron-sized grains permitted the absorption cut-off edge to be extended from 537 to 545nm. The paucity of boundaries in the films reduced the defect numbers. The response wavelengths, of a self-driven zero-biased photodetector which was based upon the film, ranged from 330 to 600nm. Micron-grained films with an area of 10cm x 10cm, and which were also prepared by heat-spraying, were stable for 1944h in air at a temperature of 298K and a humidity of 40%.

The addition of a small amount of CsI to mixed-cation halide perovskite film using a one-step method is a very good approach, but the one-step method usually generally involves an antisolvent washing process which is hard to control and unsuitable for treating large areas. By using CsF, cesium can be incorporated into a perovskite film by using a 2-step method[85]. The CsF diffuses into PbI_2 seed film and greatly increases perovskite crystallization, thus leading to high-quality films having a very long (1413ns) photoluminescence carrier lifetime, a marked stability in light, and resistance to heat and humidity. The power-conversion efficiency can be over 21%, and samples remain stable: 96% of the original efficiency remaining after 300h at 60C.

Formamidinium-Doped Lead-Halide Perovskites

Following the success of methylammonium lead bromide perovskites, nanoparticles were created[86] which were based upon formamidinium lead bromide, $CH(NH_2)_2PbBr_3$. The photophysical properties of this material could be easily varied by replacing the organic cation. Light-emitting electrochemical cells which were based upon perovskite nanoparticles could also be created by means of spray-coating. A stable luminance of 1 to $2cd/m^2$, using low driving-currents, could be obtained.

Monolithic all-perovskite tandem solar cells, with a power-conversion efficiency of 19.1%, demonstrated the improved thermal and atmospheric stability of the $[CH(NH_2)_2]_{0.75}Cs_{0.25}Sn_{0.5}Pb_{0.5}I_3$ perovskite which was used as the low gap absorber[87]. The use of uniform thick tin-lead perovskites permitted the two-terminal tandem device to attain quantum efficiencies which were greater than 80% in the near-infrared. Post-processing of the as-deposited perovskite films with methylammonium chloride vapour increased the grain sizes to more than 1μm, and increased the solar-cell open-circuit voltage and the fill-factor. The non-encapsulated solar cell retained its full potential after 150h at 85C in air.

Magnesium-doped compact TiO_2 was used as seed material for the hydrothermal growth of one-dimensional TiO_2 nano-rod arrays[88] to be combined with a $Cs_{0.05}\{(CH_3NH_3)_{0.17}[CH(NH_2)_2]_{0.83}\}_{0.95}Pb(I_{0.83}Br_{0.17})_3$ perovskite as a light absorber. With the aid of the seed layers, magnesium- and erbium-doped rutile nanorod arrays were prepared by using tetrabutyltitanate and $Er(NO_3)_3$ as the titanium and erbium precursors. Uniform straight and vertical TiO_2 nanorods, with a high areal density, were formed by means of magnesium and erbium co-modification. This improved the pore-filling and crystallization of the perovskite, aided charge-separation and suppressed recombination at the perovskite/titania nanorod interface of the solar cell. A shorter photoluminescence decay-time in the presence of magnesium/erbium doping, as compared with the doping-free case was attributed to a superior electron-extraction from the mixed-cation perovskite film. An unmodified device had an average power-conversion efficiency of 17.10%. Under the same preparation conditions, doping with magnesium, erbium or both, increased the average power-conversion efficiency to 17.54, 18.41 and 18.99%, respectively. The best device, based upon Mg/Er-modified TiO_2 nanorod arrays, had a power-conversion efficiency of 19.11%. This was a 10.33% improvement over that (17.32%) of solar cells which were based upon unmodified TiO_2 nanorod arrays.

Thin films of the methylammonium lead halides, ($CH_3NH_3PbI_3$, $CH_3NH_3PbBr_3$, $CH_3NH_3PbBr_2I$ and $CH_3NH_3PbBrI_2$), of the formamidinium lead halides, $CH(NH_2)_2PbI_3$, $CH(NH_2)_2PbBr_3$ and $CH(NH_2)_2PbBr_2I$ and of the formamidinium cesium halides, $[CH(NH_2)_2]_{0.85}Cs_{0.15}PbI_3$, $[CH(NH_2)_2]_{0.85}Cs_{0.15}PbBrI_2$ and $[CH(NH_2)_2]_{0.85}Cs_{0.15}Pb(Br_{0.4}I_{0.6})_3$ have been studied[89] with regard to their refractive index and dielectric properties within the photon energy range of 0.7 to 6.5eV. The onset of absorption near to the band-gap, together with the critical point energies and optical band transitions, shifted to higher energies as the bromine concentration in the films increased.

An easy method has been proposed[90] for constructing a p-type graded heterojunction in perovskite solar cells by using a graded distribution of hole-transporting materials such as poly[bis(4-phenyl)(2,4,6-trimethylphenyl)amine] in the shallow perovskite layer. Formation of such a structure facilitates charge-transfer and collection and passivates interfacial trap states, leading to a power-conversion efficiency of 20.05% and a steady output efficiency of 19.3% for cesium formamidinium lead halide solar cells. Non-encapsulated devices, based upon Cs–CH(NH_2)_2 lead halide perovskites, exhibited long-term stability, with more than 95% of the initial power-conversion efficiency being retained after 1440h of storage under ambient conditions.

The 2-hydroxyethylamine cation has been introduced[91] into $Cs/CH(NH_2)_2$ mixed-cation 3-dimensional perovskites in order to form stable mixed-dimensional perovskite structures having the generic form, $(2\text{-hydroxyethylamine})_2\{Cs_{0.1}[CH(NH_2)_2]_{0.9}\}_{x-1}Pb_xI_{3x+1}$. A $CH(NH_2)_2PbBr_3$ component was also used in order to improve the quality of the thin films. These mixed-dimension perovskite films exhibited better crystallization, excellent optical properties and a uniform morphology having fewer grain boundaries. Inheriting the advantages of the high-quality $CH(NH_2)_2$-based 3-dimensional perovskites, solar cells made from the mixed-dimension perovskites naturally exhibited power-conversion efficiencies which could be as high as 19.84%. The mixed-dimension cells also exhibited good long-term stability when exposed to heat and moisture. After aging in the dark at 85C, under a relative humidity of about 10% for 400h, or at 25C under a relative humidity of 55% in the dark for 2160h, non-encapsulated devices could retain 82 and 87%, respectively, of the original power-conversion efficiency.

The hybrid perovskite, $(C_4H_{12}N)_x\{[CH(NH_2)_2]_{0.83}Cs_{0.17}\}_{1-x}Pb(I_{0.6}Br_{0.4})_3$, has been investigated[92] for x-values ranging from 0 to 0.8. Small amounts of $C_4H_{12}N$ helped to improve the crystallinity, and possibly passivated the grain boundaries, thus reducing trap-mediated charge-carrier recombination and enhancing the charge-carrier mobilities. The addition of too much $C_4H_{12}N$ led to poor crystallinity and to inhomogeneous film-formation, thus greatly reducing the effective charge-carrier mobility. At low $C_4H_{12}N$ contents, the reduced recombination and increased mobility led to charge-carrier diffusion lengths of up to 7.7μm when x was equal to 0.167.

A simple single-step ligand-mediated method, involving various precursor ratios, has been used[93] for the phase-controlled synthesis of $CH(NH_2)_2PbBr_3$ cube- or rod-shaped nanocrystals. These nanocrystals are found to be fundamentally different. The cubes have properties which are similar to those of bulk $CH(NH_2)_2PbBr_3$, while the nanorods have a 2-phase microstructure, with the coexistence of a typical cubic perovskite structure and a new low-symmetry monoclinic (P2/m) phase. The two-phase nanorods also exhibit a bright dual photoluminescence with peaks centred near to 490 and 530nm, and possessing the complex luminescence dynamics which are characteristic of quasi 2-dimensional perovskites. The creation of the 2-phase nanorods was attributed to proton exchange in the presence of excess $CH(NH_2)_2$ during synthesis.

First-principles calculations have been used[94] to determine the electronic structures of lithium-, sodium-, potassium- or rubidium-containing formamidinium lead halide perovskites. The conduction band was found to be supplied with electrons from energy-levels at the 2s, 3s, 4s and 5s orbitals of the alkali metal, to energy-levels at the 6p orbital of the lead atom in the perovskite crystal. Deviations of charge distribution in the perovskite crystal led to photo-induced charge generation, electron correlation and electron-lattice interactions. Chemical shifts in the perovskite lattice arose from slight perturbations of the coordination structure, with nuclear quadruple interactions based upon the electric field gradient. First-principles theoretical predictions have also been made[95] of the electronic properties of {111} twin boundaries in pure formamidinium lead iodide, $CH(NH_2)_2PbI_3$, and in mixed-ion lead halide perovskites which contained formamidinium, cesium, iodine and bromine. The {111} twin boundary is very stable in pure $CH(NH_2)_2PbI_3$, but introduces no electron- or hole-trapping states and offers a relatively weak barrier, less than 100meV, to the transport of electrons and holes. In mixed-ion perovskites, the twin boundaries act as nucleation sites for the formation of iodine- and cesium-rich secondary phases. A reduced band-gap in the segregated phase leads to hole-trapping and may tend to increase electron-hole recombination and in turn lead to a reduced open-circuit voltage in solar cells.

An easy room-temperature method has been described[96] for preparing highly luminescent variable-colour formamidinium lead halide perovskite quantum-dots, $CH(NH_2)_2PbX_3$, where X was chlorine, bromine, iodine, chlorine/bromine or bromine/iodine. The photoluminescence emission spectra could be easily varied from 430 to 690nm. The photoluminescence spectra of $CH(NH_2)_2PbX_3$ quantum-dots included narrow emission line-widths of 16 to 34nm, with quantum-yields as high as 88% and radiative lifetimes ranging from 19.8 to 53.0ns. The $CH(NH_2)_2PbBr_3$ quantum-dots exhibited single-decay dynamics. Green-emitting diodes which were based upon the perovskite quantum-dots possessed a luminance of 403cd/m^2, and an external quantum efficiency of 2.8%. Light-emitting diodes which were based upon $CH(NH_2)_2PbBr_3$ quantum-dots exhibited a good stability of the emission peaks with increasing applied voltage.

Tetrafluoroborate (BF^{4-}) anion substitution has been used[97] to improve the properties of formamidinium-methylammonium perovskite films: $[CH(NH_2)_2PbI_3]_{0.83}(CH_3NH_3PbBr_3)_{0.17}$. The BF^{4-} incorporation leads to lattice relaxation and to longer photoluminescence lifetimes, a higher recombination resistance and a 1 or 2 orders-of-magnitude lower trap density. This in turn leads to increased power-conversion efficiencies, from 17.55 to 20.16%, plus an increased open-circuit voltage and fill-factor.

A first-principles and experimental investigation of the electronic and optical properties of perovskites revealed[98] that $[CH(NH_2)_2]_{0.85}Cs_{0.15}PbI_{2.9}Br_{0.1}$ contains a very high density of low-energy excitons, and that this density increases with increasing temperature; even at room temperature. This was attributed to strong unscreened electron-electron and partially-screened electron-hole interactions.

The effect of incorporating less than 10% of the phenethylammonium cation into $[CH(NH_2)_2PbI_3]_{0.85}(CH_3NH_3PbBr_3)_{0.15}$ is to change the film morphology and passivate grain boundaries[99]. The passivation leads to an increase in the photoluminescence intensity, carrier lifetime and open-circuit voltage of devices when the additions are smaller than 4.5%. The presence of higher-band-gap quasi 2-dimensional phenethylammonium-containing perovskite explains the grain-boundary passivation, and the quasi 2-dimensional perovskites also congregate near to the TiO_2 layer. The phenethylammonium incorporation is moreover effective in slowing the degradation of non-encapsulated devices.

Perovskite solar cells have been prepared by using mesoporous TiO_2 as an electron-transport layer, and 2,2',7,7'-tetrakis-(N,N-di-4-methoxyphenylamino)-9,9'-spirobifluorene as a hole-transport layer. The additive effects of formamidinium, rubidium, chlorine and bromine in methylammonium lead halide perovskite crystals were investigated[100]. Slight additions of $CH(NH_2)_2Cl$ and RbBr to the $CH_3NH_3PbI_3$ crystal produced homogeneous microstructures with dispersed crystal domains which improved the photovoltaic performance. An excess of added chlorine caused nanorod-like crystals to appear and impair the photovoltaic behaviour.

The methylammonium or iodide ions in $CH_3NH_3PbI_3$-based solar cells have been replaced with 20mol% of ethylammonium, formamidinium, bromine or chlorine, and the stability of the material was then determined[101] by exposure to a relative humidity of 85% or to illumination in air. Such conditions were known to cause rapid decomposition of $CH_3NH_3PbI_3$. The addition of formamidinium cations imparted the best moisture resistance and the thus-modified perovskite exhibited the lowest photochemical reactivity. The improved stability of the formamidinium-doped perovskite was attributed to the effect of the more delocalized positive charge on the formamidinium cation.

Nano-indentation and frequency-domain thermoreflectance methods were used[102] to determine the Young's moduli and thermal conductivities of 2 series of lead halide perovskites: $APbBr_3$, where A was methylammonium, formamidinium or cesium, and $CH_3NH_3PbX_3$, where X was iodine, bromine or chlorine. The room-temperature thermal conductivity tended to parallel the speed of sound in the material, while the average mean free path of phonons in the perovskites were similar. This suggests that the difference in the speed of sound, rather than the scattering time, is more important with regard to heat transfer in these materials at room temperature. Each crystal phase of the perovskites also exhibited a different trend in the temperature-dependence of the thermal conductivity. Such a phase-dependence of the thermal conductivity is slightly anomalous.

High-luminescence air-stable formamidinium lead halide perovskite quantum-dots can be prepared[103] by using high melting-point ligands. The emission spectra can be easily varied over the visible spectral range of 409 to 817nm. The photoluminescence of $CH(NH_2)_2PbBr_3$ nanocrystals has emission line-widths of 21 to 34nm, quantum-yields of up to 88% and a lifetime of 54.6 to 68.6ns; with the material remaining stable for months. Highly efficient formamidinium lead halide perovskite quantum-dot based green-emitting diodes can be made which have a luminance of $33993cd/m_2$, a current efficiency of 20.3cd/A and a maximum external quantum efficiency of 4.07%.

'Cation-tailoring' of 3-dimensional perovskites can lead[104] to reduced degradation. Two-dimensional Ruddlesden-Popper layered perovskites meanwhile exhibit superior stability but tend to make inefficient solar cells. The introduction of n-butylammonium cations into a $[CH(NH_2)_2]_{0.83}Cs_{0.17}Pb(I_xBr_{1-x})_3$ 3-dimensional perovskite leads to the formation of 2-dimensional perovskite platelets interspersed among highly-oriented 3-dimensional perovskite grains which suppress non-radiative charge recombination. Solar cells having the optimum butylammonium content exhibited an average stabilized power-conversion efficiency of 17.53% in the case of a 1.61eV band-gap perovskite and of 15.8% in the case of a 1.72eV band-gap perovskite. The stability in simulated sunlight was also increased, with 80% of their original efficiency remaining after 1000h ... or after nearly 4000h when encapsulated.

Because the excellent power-conversion efficiencies of organic lead halide perovskites originate in efficient photo-excitation and charge-carrier transport, density functional theory calculations were made[105] of formamidinium lead iodide - an alternative to methylammonium lead iodide – in order to predict its electronic structure and density-of-state characteristics. The calculated band-gap energy was 1.307eV, and the valence band was predominantly formed by the 5p-orbitals of iodine anions while the conduction band was predominantly formed by the 6p-orbitals of Pb^{2+} cations. The density-of-states of the valence band of this perovskite appeared to be smaller than in the case of methylammonium lead iodide perovskite, and this was suggested to explain the smaller power-conversion efficiencies in perovskite solar cells containing formamidinium cations.

The formation of a dense uniform thin layer on substrates is essential in the construction of perovskite solar cells which contain formamidinium plus several cations and halide anions. Any concentration of defect states reduces cell performance by decreasing the open-circuit voltage and short-circuit current-density, but the introduction of additional iodide ions into the organic-cation solution used to form perovskite layers by intramolecular exchange decreases the concentration of deep-level defects. Control of the defects in thin perovskite layers has permitted[106] the creation of solar cells having a power-conversion efficiency of 22.1% in the case of small cells, and 19.7% in the case of $1 cm^2$ cells.

The methylammonium cations in $CH_3NH_3PbX_3$ nanocrystals can be replaced by formamidinium cations via a solid-liquid-solid cation-exchange reaction. By using this type of reaction, formamidinium lead halide nanocrystals having various halide compositions could be prepared[107] by altering the starting material so as to create $CH(NH_2)_2PbX_3$ hybrid nanocrystals with emissions ranging from 395 to 700nm.

First-principles methods have been used[108] to examine the structural, electronic and optical properties of mixed formamidinium lead halide perovskites, $CH(NH_2)_2PbI_xCl_{3-x}$, where x ranged from 0 to 3. The $CH(NH_2)_2$ cations lay along the [001] direction in the trigonal $CH(NH_2)_2PbX_3$ structure, although the direction was slightly shifted due to the distortion of the PbX_6 octahedra in the mixed $CH(NH_2)_2PbI_xCl_{3-x}$ phase. The Pb-I bond-distances of 0.315 to 0.334nm were larger than the Pb-Cl bond-distances of 0.282 to 0.302 nm. With increasing I/Cl ratio, the lattice parameters and volumes of $CH(NH_2)_2PbI_xCl_{3-x}$ increased. The $CH(NH_2)_2$ cations were essential to the balancing of the crystal structure, but did not participate directly in frontier orbital transitions. They simply played the role of charge donors which contributed about 0.76e to the PbI_3 framework. The $CH(NH_2)_2PbI_xCl_{3-x}$ were direct-bandgap semiconductors, with direct-bandgap nature at the Z (0,0,0.5) symmetry-point. The valence-band maximum comprised antibonding orbitals of iodine 5p-orbitals, chlorine 3p-orbitals and a few lead 6s orbitals, while the conduction-band minimum comprised 6p lead orbitals. A combined covalent-ionic bonding mechanism operated between the lead, and iodine or chlorine ions. As the I/Cl ratio increased, the band-gap decreased and the absorption spectrum was red-shifted. The $CH(NH_2)_2PbI_3$ had an ideal band-gap of 1.53eV, and exhibited a superior absorption spectrum; particularly between 300 and 500nm.

When formamidinium was used to replace methylammonium in polycrystalline perovskite films[109], the latter underwent great structural disorder and there occurred the creation of a high density of traps. A study of the optical and electrical properties of $CH(NH_2)_2PbX_3$ single crystals (where X was bromine or iodine) showed that they exhibited appreciably increased transport properties, as compared with those of either $CH(NH_2)_2PbX_3$ polycrystalline films or $CH_3NH_3PbX_3$ single crystals. The $CH(NH_2)_2PbBr_3$ crystals, in particular, had a 5-fold longer carrier lifetime and a 10-fold lower dark carrier concentration than those of $CH_3NH_3PbBr_3$ single crystals. Carrier diffusion-lengths of 6.6μm were found for $CH(NH_2)_2PbI_3$ and of 19.0μm for $CH(NH_2)_2PbBr_3$ crystals.

Kelvin-probe microscopy of the upper surface of $[CH(NH_2)_2PbI_3]_{0.85}(CH_3NH_3PbBr_3)_{0.15}$ perovskite solar cells in the dark, under both non-contact and contact mode conditions, has shown[110] that a bias-voltage, regardless of polarity, causes changes in the contact potential. These occur due to a larger accumulation of ions at the surfaces, caused by a build-up in potential resulting from the work-function difference between the perovskite and TiO_2. There can also be a temporary or permanent morphological change due to ion-migration when a bias-voltage is applied in the contact mode. Energy-dispersive X-ray data show that $CH_3NH_3^+$, $(CH(NH_2)_2^+$ and I^- ions take part in the migration. When irradiated, the ions are redistributed by the light-induced potential, thus resulting in band-bending. This in turn encourages charge-carrier collection and is more marked at grain boundaries.

Methylammonium Lead-Halide Perovskites

The lead-halide perovskites can exhibit relatively slow recombination, in spite of their usually being prepared from solution at ambient temperatures, and that slow recombination permits the occurrence of high open-circuit voltages when they are used in solar cells[111].

Intermediate-band solar cells, with quantum-dots and a bulk semiconductor matrix, are anticipated[112] to offer a power-conversion efficiency which might well exceed the Shockley-Queisser limit. The efficiency is limited so far because the photo-absorption layers suffer from a low density of quantum dots, contain defects which are due to lattice strain and have a low band-gap energy. Solution-processed intermediate-band cells containing photo-absorption layers have been prepared in which lead-sulfide quantum dots are dispersed in methylammonium lead bromide perovskite matrices having a band-gap energy of 2.3eV. These cells exhibited two-step photon absorption via intermediate-band at room temperature by inter-subband photocurrent spectroscopy.

The X-ray photo-electron spectroscopy of methylammonium lead halide perovskite films typically reveals the presence of Pb^{II}, and sometimes of Pb^0. It is probable that the Pb^0 peaks are artefacts however, as they can be detected in films that originally contain no Pb^0. Their creation has been attributed to X-ray photolysis and to other chemical changes[113].

An extensive study has been made[114] of the components of lead halide perovskites: PbI_2 and methylammonium iodide. It focused particularly on the lead-iodine chemical bond in the PbI_2 precursor. It was concluded that the inclusion of spin-orbit coupling details was essential when performing density functional theory computations.

Dielectric losses in the GHz regime in methylammonium-based materials are dominated by the rotational dynamics of the organic cation. As the temperature is increased from 160 to 300K, the rotational relaxation time decreases from 400 to 6ps in $CH_3NH_3PbI_3$ and from 200 to 1ps in $CH_3NH_3PbBr_3$. Negligible temperature-dependent variations in the rotational relaxation time occur in the case of $[CH(NH_2)_2]_{0.85}(CH_3NH_3)_{0.15}Pb(I_{0.85}Br_{0.15})_3$. It was concluded[115] that dipolar reorientation of the CH_3NH_3 cation did not affect the charge-carrier mobility and lifetime in metal halide perovskites, and that charge-carriers did not sense the relatively slow-moving CH_3NH_3 component.

A model which accounts for phonon-assisted free-exciton and free-carrier trapping in $CH_3NH_3PbI_{3-x}Cl_x$ has been developed[116]. Optical spectroscopy detected an appreciable co-existence of the tetragonal and orthorhombic phases at low temperatures, and permitted an estimate to be made of the longitudinal optical phonon energy, exciton binding energy and temperature-dependent electronic band-gap. These quantities could then be used to model the temperature-dependent and fluence-dependent time-resolved photoluminescence decays, and thus permit a demonstration of how shallow traps, from which carriers can be re-excited, could account for delayed recombination in the lead halide perovskites. The trap-state density attained a maximum value at the tetragonal-to-orthorhombic phase transition at about 140K. This implied the formation of disorder-induced trap states which dominated the recombination dynamics in $CH_3NH_3PbI_{3-x}Cl_x$.

When $CH_3NH_3PbBr_3$ and $CH_3NH_3PbI_3$ are poured onto glass slides, and physically combined, they can undergo halide-exchange and form mixed-halide films. A change in the composition of 130nm-thick films occurs as Br^- diffuses toward the $CH_3NH_3PbI_3$ film and I^- diffuses toward the $CH_3NH_3PbBr_3$ film. Their interdiffusion can be monitored by noting changes in absorption, and this permits the direct measurement of thermally-activated halide diffusion in perovskite films. An increase in the rate-constant of halide diffusion, from 8.3×10^{-6}/s at 23C to 3.7×10^{-4}/s at 140C, obeyed an Arrhenius relationship with an activation energy of 51kJ/mol[117].

As well as being excellent solar-cell materials, these perovskites also exhibit a surprising propensity to serve as memory-storage devices. Vacancies in lead halide perovskite nanostructures can be used to create nano floating gate memories[118]. That is, $CH_3NH_3PbBr_3$ nanocrystals have been evenly spread onto a CdS nanoribbon by dip-coating so as to form a CdS-nanoribbon|$CH_3NH_3PbBr_3$-nanocrystal structure. Due to the existence of adequate carrier-trapping states in the $CH_3NH_3PbBr_3$ nanocrystals, the device had a memory-window of up to 77.4V, a retention-time of 12000s, a current ON/OFF ratio of 7×10^7 and a long-term atmospheric stability of 50 days. A resistive storage device has been described[119] which is of the form: $CH_3NH_3PbI_{3-x}Cl_x$|F-doped SnO_2. This stores information as 2 levels of resistance-state, as induced by electrical-probe excitation. The perovskite layer is deposited onto a F-doped SnO_2-coated glass by single-step solution spin-coating in air. When the probe was silver, the device exhibited bipolar resistive switching behavior with a 10^6 ON/OFF resistance ratio. The memory-cell had a minimum endurance of 10^4 cycles and a minimum retention time of 2 x 10^3s.

A study of lead halide based transparent photovoltaics, which exhibited both increased efficiency and transparency, examined[120] the effects of halide layer-thickness and composition. This showed that lead halide transparent photovoltaics endowed with ultraviolet-wavelength selective absorption could attain power-conversion efficiencies of more than 1% when the average visible transmission was over 70%.

First-principles determinations involving a many-species expansion have been made[121] of the contributions of the various atoms and orbitals to the electronic and optical properties of chlorine, bromine and iodine lead-halide perovskites. This permitted the atomic and orbital contributions to be quantified as a photon-energy function.

The addition of large hydrophobic cations to these perovskites has markedly improved the environmental stability of photovoltaic cells which are based upon them, but this can lead to the formation of 2-dimensional structures within the material and a consequential dielectric confinement, higher exciton binding energies, wider band-gaps and limited charge-carrier mobility. These effects are not detrimental however to charge transport in suitably processed films which include a self-assembled thin layer of quasi 2-dimensional perovskite, interfaced with a 3-dimensional $CH_3NH_3PbI_3$ perovskite layer. Time-resolved photoluminescence and photoconductivity spectra reveal[122] the occurrence of charge-carrier recombination and transport through the film when the quasi 2-dimensional or 3-dimensional layers are selectively excited. Modeling of these dynamics showed that, while the charge-carrier mobility was lower within the quasi 2-dimensional region, charge-carrier diffusion to the 3-dimensional phase led to a rapid recovery in photoconductivity even when the quasi 2-dimensional region was initially photo-excited. A blue-shifted emission which originated from the quasi 2-dimensional regions appreciably overlapped the absorption spectrum of the 3-dimensional perovskite. This permitted the occurrence of so-called 'heterogeneous photon recycling', which was deemed to be a highly effective process.

The organic compound, 1-(ammonium acetyl)pyrene, has been used[123] to prepare 2-dimensional/3-dimensional $CH_3NH_3PbI_3$ heterostructures in which the high humidity resistance and strong absorption in the ultra-violet region of the pyrene group imparts considerable stability. There was no degradation, in a relative humidity of about 60%, even after 6 months. There was also a high ultraviolet-radiation stability, with hardly any degradation after 1h of ultra-violet ozone treatment. The maximum efficiency of treated $CH_3NH_3PbI_3$ perovskite solar cells was 14.7%, with essentially no hysteresis.

Photoluminescence spectroscopy has been used[124] to determine the effect of introducing heterovalent Bi^{3+} impurities into monocrystalline $CH_3NH_3PbX_3$ and $CH(NH_2)_2PbX_3$, where X was iodine or bromine. Elemental analysis showed that the Bi^{3+} ions were more easily incorporated into $CH_3NH_3PbI_3$ than into $CH(NH_2)_2PbBr_3$ or $CH_3NH_3PbBr_3$. The perovskite single crystals exhibited a blue-shift of the room-temperature photoluminescence spectra as the Bi^{3+} content was increased. The blue-shift was attributed to a weaker photon-recycling effect that was caused by shorter photoluminescence lifetimes. The room-temperature photoluminescence spectra of $CH(NH_2)_2PbBr_3$ and $CH_3NH_3PbI_3$ were relatively insensitive to impurity-doping, as compared with the photoluminescence spectrum of $CH_3NH_3PbBr_3$. Doping with Bi^{3+} did however affect the low-temperature photoluminescence of each material in a different manner; thus suggesting that this depended upon the type of ion in the host.

The interface between a NiO_x hole transport layer and a perovskite absorber layer can be modified by using a 4-bromobenzylphosphonic acid-based self-assembled monolayer, thus leading to an improved photovoltaic performance[125]. Modification of the NiO_x layer is performed by dip-coating, allowing adequate time for self-assembly. A change in the surface free-energy, and the creation of a non-polar surface occurs, together with the residual presence of phosphor and bromine. The resultant solar cells exhibit an increased photovoltage. A comparison of devices, without and with modification, shows that the photovoltage increases from 0.978 to 1.029V, with the maximum open-circuit observed being 1.099V. The increase in photovoltage then leads to an improved power-conversion efficiency.

A slowing of carrier-cooling in these perovskites could lead to the creation of efficient hot-carrier solar cells. One task is to understand the retardation of carrier relaxation, and another is to demonstrate hot-carrier harvesting. A further task is to assess the slowing effect of energy and momentum relaxation upon spontaneous and stimulated light-emission. Nanocrystals of halide perovskites are perfect for the investigation of these phenomena because of their bright emissions and high optical gain, and the carrier-cooling bottleneck is accentuated, as compared with bulk material, due to confinement. The luminescent properties of $CsPbBr_3$, $CH(NH_2)_2PbBr_3$ and $CH(NH_2)_2PbI_3$ nanocrystals have been investigated[126] in the strong photo-excitation regime. In the first 2 materials above, amplified spontaneous emission predominated over radiative recombination at average carrier occupancies per nanocrystal which were greater than 5 to 10. Under the same photo-excitation conditions, the $CH(NH_2)_2PbI_3$ nanocrystals contained a longer-lived population of hot carriers; resulting in a competition between hot luminescence, stimulated emission and defect recombination. The interplay between the various processes appeared to be affected by factors such as temperature, purity, film-morphology, excitation pulse-width and wavelength.

Certain materials have also been considered for use as solid-state lighting devices, even though their quantum-yields for white photoluminescence are only in the range of 0.5 to 9%. The quantum-efficiency of a hybrid metal halide can be greatly improved[127] however if it contains a polymorph of the $[PbX_4]^{2-}$ perovskite-type layers, where the photoluminescence quantum-yield can be 45%. The addition of various piperazines can lead to hybrid lead halides comprising either perovskite layers or post-perovskite chains which strongly affect the presence of self-trapped states for excitons.

A non-stoichiometric molar ratio of CH_3NH_3I and PbI_2 precursors in dimethyl sulphoxide solution has been used[128] to control the quality of perovskite layers. Uniform dense $CH_3NH_3PbI_3$ solar cells were created simply by changing the $CH_3NH_3PbI_3$ solubility. Homogeneous grain-sizes and regular inter-grain arrangements were also obtained in that way. Solar cells which were produced under the optimum conditions exhibited a higher current density and power-conversion efficiency; attributed to a Lamer-model effect on internal light absorption.

Density functional theory estimates, in the generalized gradient approximation, were made[129] of the band-gaps, densities of states, absorption coefficients, refractive indices, dielectric constants and elastic constants of orthorhombic methylammonium lead halide perovskites. The stiffness of $CH_3NH_3PbX_3$, with X = I, Br or Cl, was investigated by calculating the Young's modulus and among these, $CH_3NH_3PbI_3$ was the stiffest, with E = 57.24GPa. The family members were characterized by their band-gaps, with $CH_3NH_3PbI_3$, $CH_3NH_3PbBr_3$ and $CH_3NH_3PbCl_3$ having values of 1.626, 2.207 and 2.748eV, respectively. They also exhibited remarkable absorption coefficients over the visible spectrum. There was an anisotropy of the optical properties in the near- and mid-ultraviolet band.

The solar cells which use SnO_2 as the electron-transport layer improve as a result of light-soaking. Photoluminescence and electrochemical impedance spectroscopy data show[130] that the improvement is due to a reduced non-radiative recombination, together with a reduction of the extrinsic electron concentration in the oxide. The enhancement could also be brought about by exposing devices to a high vacuum at ambient temperatures. It was suggested that this improvement resulted from the desorption of hydrogen from oxygen vacancies in the oxide. Gallium-doped SnO_2-based devices exhibit a lesser light-soaking effect and contain fewer oxygen vacancies. It was concluded that high extrinsic electron concentrations in SnO_2 are undesirable because of the role which they play in non-radiative recombination. The reduction in electron density which occurs when SnO_2 is incorporated into a perovskite diode therefore improves the cell's performance.

Films of $CH_3NH_3PbI_3$ and $(CH_3NH_3)_{1-x}[CH(NH_2)_2]_xPbI_3$ mixed-cation perovskites were produced[131] by using methylammonium acetate precursor and chlorobenzene; thus provoking a synergistic effect. Methylammonium acetate led to the crystallization of (112)/(200), rather than (110) orientations, while the chlorobenzene tended to accelerate the crystallization of $CH_3NH_3PbI_3$ (112)/(200). Meanwhile, (110) and (112)/(200) phases coexisted in the case of the $(CH_3NH_3)_{1-x}[CH(NH_2)_2]_xPbI_3$ mixed-cation perovskite and $CH(NH_2)_2^+$ could promote crystallization of the (110) orientation alone. Chlorobenzene further promoted sole growth of the (110) orientation. Optimization of the process, using 16% of methylammonium acetate, led to a $(CH_3NH_3)_{0.9}[CH(NH_2)_2]_{0.1}PbI_3$ solar cell having an efficiency as high as 17.5%.

The temperature-dependence of the Raman bands of $CH_3NH_3PbX_3$ (X = I, Br) was measured[132] at 100 to 290K for X = Br and at 110 to 340K for X = I. Broad bands at about 326/cm in the former case and at about 240/cm in the latter case were attributed to CH_3NH_3–PbX_3 cage-vibrations. These bands exhibited an anomalous temperature dependence which was attributed to a motional narrowing that originated in rapid changes between the orientational states of $CH_3NH_3^+$ in the cage. The phase transitions were characterized by changes in the bandwidths and peak positions of the CH_3NH_3–cage vibration and by some bands which were associated with the NH_3^+ group.

It has been demonstrated that the crystalline orientation of a perovskite film can play an important role in moisture degradation[133]. Films which were made in different ways exhibited differing degradation behaviors, and this divergence was traced to differences in the crystallographic characteristics. Films which were prepared using a single-step $PbCl_3$-precursor method suffered much more rapid degradation than did films which were prepared using single-step acetate-precursor based or 2-step methods. The resultant film was always $CH_3NH_3PbI_3$ but the orientations were different. An orientation which displayed many (100) and (200) planes exhibited the formation of a hydrated phase and this led to much faster decay. An orientation which exhibited many (110) and (220) planes underwent slow decomposition without involving any intermediate hydrated phase at relative humidities of 70 to 80%. There was the possible formation of a hydrate on the (110) plane at relative humidities of about 90%. Large quantities of water are of course detrimental to lead halide perovskite solar cells; due mainly to a resultant decomposition of the perovskite layer. On the other hand, small quantities of water play a pivotal role in the crystallization of the perovskite and in the behaviour of related devices. Impedance spectroscopy has been used[134] to analyze the changes in the electronic properties of methylammonium lead tri-iodide which are produced by environmental humidity. This showed that water which was absorbed from the environment caused a huge increase in the capacitance; reaching levels as high as the accumulated capacitance found in devices and which produces the large hysteresis between forward and reverse J-V curves. Together with the striking increase in capacitance, water absorption also produced an appreciable increase in the conductivity. An activation energy of 0.52eV was deduced for the electronic transport; a value comparable to the activation energy of ionic transport. This suggested that ambipolar diffusion was the link between the various phenomena. It has been shown[135] that through-thickness cracks, produced by bending-tension in thin films of fragile methylammonium lead tri-iodide and formamidinium lead tri-iodide, could be easily repaired by exerting a slight compressive stress at room temperature or by heat treating at moderate temperatures. The repair was time-dependent, showing that simple mass transport was an important part of the process.

Efficient solar cells depend upon good charge-extraction at the interface between the perovskite and the charge-transport layer. The presence of under-coordinated metal or halogen ions at the interface can impair this process. The problem can be reduced[136] by introducing an interfacial small-molecule anchoring layer which is based upon dimethylbiguanide. Interactions between the nitrogen-containing functional groups of the latter material, and unsaturated ions in $CH_3NH_3PbI_3$ perovskites mean that electron-extraction by TiO_2 is markedly improved. The power-conversion efficiency of $CH_3NH_3PbI_3$ solar cells then increases from 17.14 to 19.1%; together with a much reduced hysteresis.

Truxene derivatives are promising candidate materials for the passivation of defects by deposition onto hybrid lead halide perovskite thin films. The semiconducting properties can be varied by modifying the chemical structure thus permitting, upon light irradiation, interfacial charge-transfer between a perovskite film and truxene molecules. It has been observed[137] that these molecules dampen non-radiative carrier recombination dynamics in the perovskite thin film via supramolecular complex-formation between truxene molecules and Pb_2^+ defects at the perovskite surface. Such a supramolecular complex affects neither the carrier recombination kinetics nor carrier-collection, but introduces some hysteresis into the photocurrent versus voltage curves of solar cells.

The NaI-doping of a perovskite layer was expected[138] to modify photophysical properties and improve the performance of solar cells. The perovskite layer was prepared by one-step solution spin-coating, using various dopant concentrations of NaI. The absorption band-edge and the peak position of the photoluminescence spectrum of the doped thin film were red-shifted so that the band-gap of the semiconductor film narrowed. It was concluded that NaI-doping is an effective means for modifying the properties of perovskite films.

The perovskite thin films can be prepared by 2-step spin-coating, where PbI_2 is first deposited, followed by a methylammonium halide. The addition of isopropanol during the first step produces a more porous and less crystalline PbI_2 film. It has been noted[139] that the surface roughness of alcohol-treated PbI_2 layers is almost twice that of untreated layers. The surface of treated films was fractal-like, with the same roughness pattern being seen at different scales. Larger grain-sizes and greater crystallinity were observed in the case of treated films, regardless of the partial replacement of iodide by chloride or bromide. Perovskite solar cells which were based upon the alcohol-treated PbI_2 had average power-conversion efficiencies of 10.27%, as compared with a 7.20% efficiency without alcohol-treatment.

A unique and novel nano-structured perovskite, ethyl ammonium lead chloride ($C_2H_5NH_3PbCl_3$), was prepared via a co-precipitation route using ethylamine ($C_2H_5NH_2$) and hydrochloric acid as precursors, with an aqueous solution of $Pb(CH_3COO)_2 \bullet 3H_2O$[140]. The material was then deposited onto TiO_2 film by spin-coating[141]. The resultant perovskite-based solar cell exhibited a power-conversion efficiency of about 6.01%.

It has been shown[142] that the stability of these materials can be markedly improved by fluorination of the methylammonium cation. Density functional theory was used to predict the optimum stability of a perovskite as a result of adding low concentrations of cations. The optimum fluorination level had little effect upon the band-gap or the volumetric expansion of $CH_3NH_3PbI_3$. Fluorination had a tendency to stabilize the material, due to the strengthening of initially weak hydrogen bonds between $CH_3NH_3^+$ cations and the surrounding lead-iodide framework. The strengthening resulted from internal structural deformations which were related to the formation of long C-N bonds. Fluorination was also expected to reduce iodine-vacancy mediated diffusion under a bias voltage.

It may also be possible to prepare these materials via solidification of a thin layer of the melt. It is possible to vary the melting point of layered hybrid lead iodide perovskites by more than 100C by modifying the structures of alkylammonium-derived organic cations. By introducing alkyl chain-branching, and extending the length of the basic alkylammonium cation, melting points as low as 172C can be arranged. High-quality thin films of layered hybrid lead iodide perovskites can then be made by using a solvent-free process in ambient air[143].

The majority of perovskite photodetectors comprise polycrystalline film, but this is undesirable due to the presence of grain boundaries. Monocrystalline lead halide wafers of up to 10mm in size can be prepared by combining space-confined growth with seed-induced crystallization[144]. The use of space-confined growth by itself produces multiple crystals which are smaller than 3mm. Because of the absence of grain boundaries, the trap-density of large monocrystalline perovskite wafers is only $2.36 \times 10^{10}/cm^3$. This is much lower than that of polycrystalline films. Photodetectors which were based upon monocrystalline perovskite wafers exhibited a light ON/OFF current ratio of 4.3×10^3, a response time of 770µs and a linear dynamic range of 119dB. The lowest detectable illumination power density was $80nW/cm^2$. The photodetector here exhibited the highest external quantum efficiency of 904%, a responsivity of 3.87A/W and a detectivity of 1.77 $\times 10^{13}$. The monocrystalline perovskite devices were also stable, and retained more than 80% of the original photocurrent after 720h of storage in air.

Like silicon, $CH_3NH_3PbI_3$ has an anomalously large Auger coefficient, with the theoretical external luminescence efficiency decreasing to about 95% under open-circuit conditions. The effect is reduced at the operating point, where the carrier-density is lower and leads to a strangely high fill-factor of some 90.4%[145]. This compensates somewhat for the Auger penalty, and produces an operating-point voltage and power-conversion efficiency which are close to being ideal for the $CH_3NH_3PbI_3$ band-gap.

Hybrid organic-inorganic perovskites which are subjected to heating when serving as solar-cells exhibit an anomalous increase in the charge-carrier lifetime at the higher temperatures. This unusual but useful phenomenon has been explained[146] by using time-dependent density functional and non-adiabatic molecular dynamics methods. In the case of tetragonal $CH_3NH_3PbI_3$, the temperature had differing effects upon the organic and inorganic sub-systems and this led to subtle structural changes. Charge-phonon interactions decreased because libration of the organic component at higher temperature reduced the oscillation amplitude of the Pb-I lattice which hosted electrons and holes. Thermal disorder localized wave-functions and reduced non-adiabatic charge-phonon coupling. Tilting of the inorganic octahedra widened the band-gap, thus extending charge-carrier lifetimes still further.

A correlation exists[147] between the photoconductivity and the photoluminescence, expressed as functions of the incident photon flux, in hybrid and all-inorganic perovskite single crystals. A universal scaling behavior has been detected during simultaneous measurements of the Hall effect, photoconductivity and photoluminescence. The photoconductivity exhibits a cross-over, in the power-law dependence, between an exponent of unity and one of ½. The photoluminescence meanwhile exhibits a cross-over between an exponent of 2 and one of 3/2. The correlation appears to be independent of the cation-type and of the crystallographic phase.

Methylammonium lead halide perovskites of the form, $CH_3NH_3PbX_3$ (X = Cl, Br or I), have been studied[148] with regard to their use as the active medium in luminescent solar concentrators, It was found that the $CH_3NH_3PbCl_3$ perovskite, with a 2.0wt% perovskite content in polyvinyl alcohol, led to the highest efficiency and least dissipative effect.

It has been suggested that slow radiative recombination, due to a slightly-indirect band-gap, explains the high efficiency of lead halide perovskite solar cells, and first-principles calculations have been made[149] of the radiative recombination rate of $CH_3NH_3PbI_3$. Given that the CH_3NH_3 molecule rotates, even at room temperature, the momentum mismatch between the band edges was determined as a function of the orientation of the CH_3NH_3 molecule. It was found that the indirect nature of the band-gap suppressed the radiative recombination rate by less than a factor of 2, and that the radiative recombination coefficient was as high as that for direct-gap semiconductors.

The role played by spin-orbit coupling in the non-radiative relaxation of hot electrons and holes in $CH_3NH_3PbI_3$ was studied[150] by using a non-adiabatic molecular dynamics method involving 2-component spinor wave functions that were solutions to the relativistic Kohn-Sham equations. This showed that spin-orbit coupling increased the contributions made by lead orbitals to the conduction and valence bands. The Kohn-Sham orbitals thereby became more sensitive to nuclear motions and this led to increased non-adiabatic coupling. Spin-orbit coupling thus markedly speeded-up the electron and hole relaxations. It was proposed that the fast hot-carrier relaxation, enabled by spin-orbit coupling, led to a rapid transition into the long-lived triplet state that extended charge-carrier life and explained the high efficiency of perovskite solar cells. The rapid relaxation of above-bandgap hot-carriers in fact places a limit on the efficiency of the cells. Hot-carrier cooling may be sensitive to the composition, as well as the energy and density of the excited states. Sequences of ultra-fast optical pulses have been used[151] to compare directly the intraband cooling dynamics of $CH(NH_2)_2PbI_3$, $CH(NH_2)_2PbBr_3$, $CH_3NH_3PbI_3$, $CH_3NH_3PbBr_3$ and $CsPbBr_3$. An approximately 100 to 900fs cooling-time, with slower cooling at higher hot-carrier densities, was detected. The effect was strongest in the case of all-inorganic cesium-based materials, as compared with hybrids involving organic cations. These data helped to identify the cause of the so-called hot-phonon bottleneck in lead-halide perovskites, as well as a thermodynamic contribution of symmetry-breaking organic cations to rapid hot-carrier cooling.

It has been found[152] that, under illumination, there occurs a reversible generation of paramagnetic Pb_3^+ defects in $CH_3NH_3PbI_3$ when prepared under ambient conditions. The effect is caused by the presence of Pb-O defects, in the perovskite structure, that can trap photo-generated holes; perhaps via simultaneous associated oxidation and ion-migration. It was assumed that one charge was trapped per each new paramagnetic center, meaning that there was no nett contribution to the photocurrent.

It is noted[153] that terahertz thin-film total internal reflection spectroscopy exhibits a higher sensitivity to the vibrational properties of thin films, as compared with that of THz transmission spectroscopy. This extra sensitivity can be used, for example, to monitor photo-induced changes to the structure of $CH_3NH_3PbI_{3-x}Br_x$ (x = 0, 0.5, 1 or 3) thin films. Illumination first strengthens the phonon modes, around 2THz, which are associated with Pb-I stretching modes coupled to CH_3NH_3 ions, while the 1THz twisting modes of inorganic octahedra are unaltered in strength. Under longer-term illumination, the 1THz phonon modes of encapsulated films slowly reduce in strength while, in films which are exposed to moisture and oxygen, the phonons weaken more rapidly and are blue-shifted in frequency.

Measurements of the intensity-dependence of time-resolved photoluminescence, in $CH_3NH_3PbI_3$ samples having various grain-sizes, showed[154] that the photoluminescence decay in all of the samples could be explained in terms of bimolecular radiative recombination and trap-assisted Shockley–Read–Hall recombination.

Devices which were built by using lead acetate as a source material were found[155] to exhibit an efficiency of about 13%. The addition of excess amounts of lead to the precursor caused it to be found in the perovskite active layer. Changes in crystallinity occurred without any significant decrease in charge-carrier lifetime, and this was attributed to the formation of PbI_2. This degradation product is known to impart beneficial effects, due to its tendency to passivate grain boundaries and alter the band-structure at the interface between the active layer and the electron-transport layer. It was noted that 5mol% of excess lead was the optimum concentration for imparting an improved efficiency and stability to devices: the latter could retain more than 50% of the original efficiency, after 1h of simulated solar exposure, when a 10mol% excess of lead was present.

When $CH_3NH_3PbI_3$ was prepared in humid air, the humidity led to the formation of a black layer on top of the dark-brown perovskite[156]. The black layer had an optical absorption tail which extended into the near-infrared region, together with poor X-ray diffraction. It also impaired solar-cell performance. Formation of such a layer could however be prevented by using a rapid-deposition crystallization process involving a suitable solvent.

Marked improvements in the opto-electronic properties of lead halide perovskites can be obtained by adding monovalent ions which have radii which are close to that of Pb^{2+}. In the case of solution-processed $CH_3NH_3PbI_3$ for example, Na^+, Cu^+ and Ag^+, can be added. These are of lower valence than Pb^{2+} but have similar radii. Unlike Ag^+ and Cu^+, the Na+ appeared mainly at grain boundaries and surfaces. Synchrotron X-ray diffraction revealed[157] a large shift in the main perovskite peaks of monovalent-cation films, while synchrotron X-ray photo-electron measurements indicated an appreciable change in the valence-band position of Cu- and Ag-doped films. The perovskite band-gap remained the same however, thus suggesting the occurrence of a shift in the Fermi-level position toward the middle of the band-gap. It was proposed that these changes implied the occurrence of a so-called de-doping effect, which led to lower energetic disorder.

The study of ionic conductivity in these materials is rather difficult because of the problem of distinguishing between ionic and electronic contributions. The preparation of a low defect-density monocrystalline $CH_3NH_3PbBr_3$ solar cell having no hole-transport layer here eliminated any electronic current. Impedance spectroscopy could then monitor the ionic diffusion and the accumulation of ions at external contacts. Diffusion coefficients were deduced from the relationship between the $CH_3NH_3PbBr_3$ thickness and the characteristic diffusion transition frequency. The diffusion coefficient of Br^- was thus found[158] to be $1.8 \times 10^{-8} cm^2/s$.

The effect of doping perovskites of the form, $[CH(NH_2)_2]_{0.83}(CH_3NH_3)_{0.17}PbBr_{0.51}I_{2.49}$, with Cs^+, Rb^+, K^+, Na^+ and their combinations was studied[159]. It was noted that Cs^+ could replace organic cations in the perovskite structure whereas Rb^+, K^+ and Na^+ did not appear to do so. Samples which were doped with potassium and sodium nevertheless had considerably longer fluorescence lifetimes.

Perovskite nanocrystals have been prepared[160], using a surfactant-free method, by introducing antimony iodide into the perovskite precursors. Surface-bound $(CH_3NH_3)_3Sb_2I_9$ layers restricted the growth of $CH_3NH_3PbI_3$, leading to nanocrystals which were about 50nm in size; with no long-chain organic ligands on the surface that could inhibit charge transport. As compared with bulk material, the nanocrystals exhibited a stronger photoluminescence.

Spiro-MeOTAD [2,2′,7,7′-tetrakis(N,N-di-p-methoxyphenylamine)-9,9′-spirobifluorene], used as a hole-transport layer in perovskite-based solar cells, often suffers from the presence of pinholes in the solution-processed films. The formation of these pinholes is attributed[161] to the presence of small amounts of secondary solvents such as water, 2-methyl-2-butene or amylene which have a low miscibility with the main solvent – such as chlorobenzene - used to dissolve the spiro-MeOTAD.

Reconstruction along the (112)/(200) planes rather than the (110)/(002) planes of $CH_3NH_3PbI_3$ can be ensured by modifying the precursor solution. Methylammonium acetate additions change the initial orientation of PbI_2 from (001) to (101) via the formation of an intermediate PbI_2 adduct, and lead to the formation of dense perovskite films comprising large grains on (112)/(200) planes[162]. These differ from the needle-like

Materials Research Forum LLC
https://doi.org/10.21741/9781644900819

crystallites which form along (110)/(002) planes in the absence of methylammonium acetate. The films possess better opto-electronic properties and resultant solar cells exhibiting efficiencies of up to 18.91% have been obtained by using solutions having a methylammonium acetate content of 11%. Some of the associated properties can be anisotropic.

Doping with potassium has been proposed[163] to be an effective means of solving the current-voltage hysteresis problem. The use of ^{39}K magic-angle scattering nuclear magnetic resonance at 21.1T has permitted the atomistic study of potassium-containing phases which form as a result of the KI-doping of multi-cation and multi-anion lead halide perovskites. There was no sign of potassium incorporation into 3-dimensional perovskite lattices, but rather the formation of a mixture of potassium-rich phases and unreacted KI. In the case of bromine-containing lead halide perovskites doped with KI, a mixture of KI and KBr resulted, with a change in the Br/I ratio of the perovskite phase. Doping with both cesium and potassium led to the formation of non-perovskite lead iodide phases.

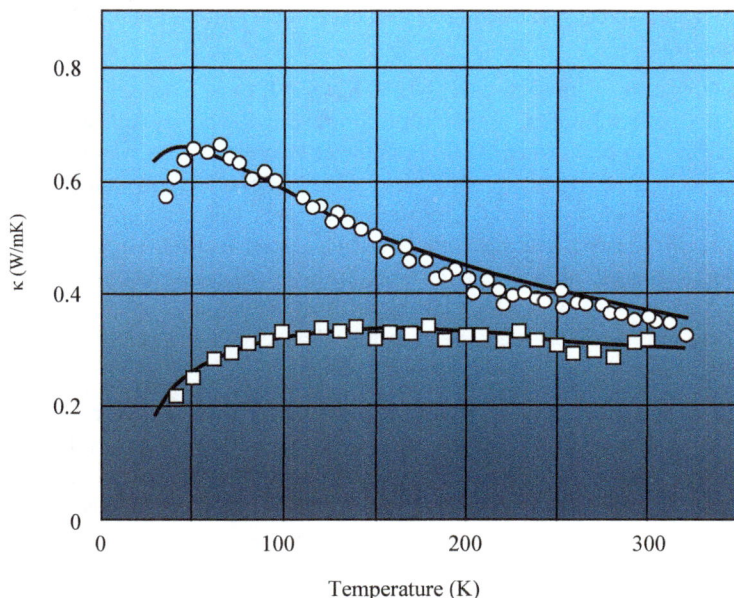

*Figure 8. Thermal conductivity of CsPbBr$_3$ (upper curve)
and CH$_3$NH$_3$PbBr$_3$ (lower curve)*

When polycrystalline $CH_3NH_3PbI_3$ films are interfaced with ferro-electric $PbZr_{0.2}Ti_{0.8}O_3$, there can be a more than 10-fold increase in the d_{33} piezoelectric coefficient. The d_{33} coefficient ranges from 0.3 to 0.4pm/V for $CH_3NH_3PbI_3$ which is deposited onto gold, indium tin oxide or $SrTiO_3$ surfaces[164]. There is a small phase-angle fluctuation at length-scales shorter than the grain size. When $CH_3NH_3PbI_3$ is deposited onto epitaxial $PbZr_{0.2}Ti_{0.8}O_3$ films, large-scale polar domains are observed having a phase-angle which is close to 180°. By distinguishing between the piezoresponse contributions of $CH_3NH_3PbI_3$ and $PbZr_{0.2}Ti_{0.8}O_3$ layers, the d_{33} coefficient of $CH_3NH_3PbI_3$ was found to be about 4pm/V. This was attributed to an increased alignment of the methylammonium molecular dipoles that was caused by the unbalanced surface potential of $PbZr_{0.2}Ti_{0.8}O_3$.

Studies of mechanisms governing thermal transport in $CH_3NH_3PbI_3$, $CH_3NH_3PbBr_3$ and $CsPbBr_3$ showed[165] that the thermal conductivity of $CH_3NH_3PbBr_3$ nanowires is lower when compared with that of $CsPbBr_3$ nanowires (figure 8). This was attributed to cation dynamic disorder. Differing temperature-dependent thermal conductivities were found for $CH_3NH_3PbBr_3$ and $CH_3NH_3PbI_3$ (figure 9). This was attributed to accelerated cation dynamics in $CH_3NH_3PbBr_3$ at low temperatures, and the combined effects of a lower phonon group-velocity and a higher Umklapp scattering-rate in $CH_3NH_3PbI_3$ at high temperatures.

When films of the perovskites, $CH_3NH_3PbX_3$ (X = I, Br or Cl), were grown on fluorine-doped tin oxide substrates, X-ray diffraction revealed good crystallization and a strong (100) diffraction peak, when X was bromine or chlorine[166]. This was shifted to a higher angle as compared with the (110) peak of $CH_3NH_3PbI_3$. The band-gaps were 1.63, 2.37 and 3.11eV for X = I, Br and Cl, respectively; energy values which closely matched the positions of the peaks deduced from photoluminescence data. A notable absorption-dip and emission-peak occurred in $CH_3NH_3PbBr_3$; suggesting the presence of greater crystallinity, in spite of the identical preparation conditions. The wave-numbers of the main infra-red frequencies decreased slightly with the ionic radius of the halogen ion, and this was attributed to an increasing polarizability.

The perovskite-based solar cells can be improved by changing the organic cation. For example, in the case of mixed methylammonium–guanidinium (GA) perovskites, $(CH_3NH_3)_{1-x}GA_xPbI_3$, the Pawley-fit method confirms the formation of a $GAPbI_3$ tetragonal phase, and up to 20% of the guanidinium cation can be incorporated into the methylammonium lead iodide perovskite[167]. This leads to a lattice enlargement which can be monitored by studying shifts in the diffraction peaks of the $CH_3NH_3PbI_3$ tetragonal lattice. The long-term stability was improved in mixed perovskites having a low guanidinium content. The band-gap was shifted to lower energies, and the absorption band-gap decreased slightly when guanidinium replaced up to 20% of the methylammonium in $CH_3NH_3PbI_3$.

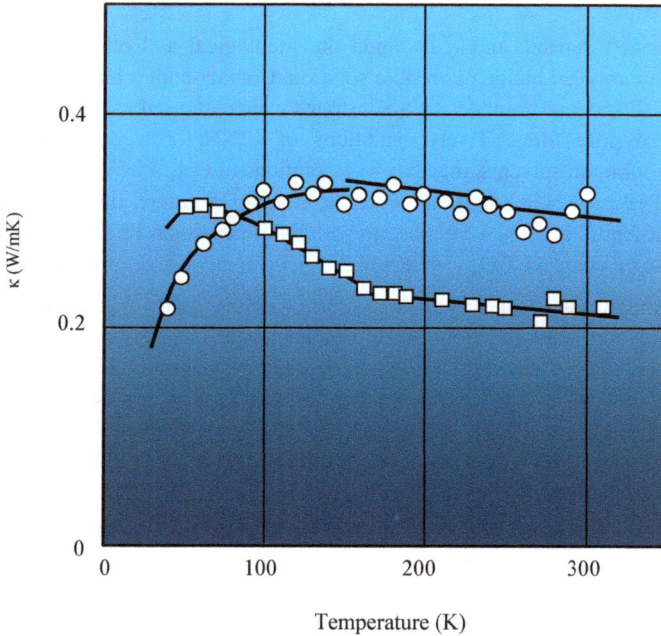

Figure 9. Thermal conductivity of $CH_3NH_3PbBr_3$ (upper curve)
and $CH_3NH_3PbI_3$ (lower curve)

In $CH_3NH_3PbI_{3-x}Cl_x$-based planar solar cells with a power-conversion efficiency of 15.8%, an illumination intensity-dependence of the current-density voltage plots revealed[168] the occurrence of trap-assisted recombination at low fluences. Capacitance-voltage spectra exhibited distinct variations over a wide range of alternating-current modulation frequencies; with and without photo-excitation. The frequency-dependent chemical capacitance in this material was related to the surface-related and bulk-related densities-of-states; as checked by fitting the corresponding density-of-states to a Gaussian distribution function. The electronic sub-gap trap-states in the solution-processed material, and their distribution, varied from the surface to the bulk. Surfaces that were next to the electron-transport and hole-transport layers had an analogous density-of-states. Photo-induced and bias-induced giant dielectric responses nevertheless occurred and a marked reduction at frequencies greater than 100kHz was attributed to dielectric loss.

The coating step depends markedly upon the rheological and other properties of the methylammonium lead halide perovskite solution. Contact-angles have been measured[169] on the PEDOT:PSS and compact TiO_2 substrates which are used as the underlying layers of most perovskite films. Twelve solutions of $CH_3NH_3PbI_3$ and $CH_3NH_3PbI_{3-x}Cl_x$ dissolved in common solvents, together with solutions of PbI_2, $PbCl_2$ and CH_3NH_3I, were examined. The perovskite solutions were Newtonian in nature. The apparent contact-angles on mesoporous TiO_2 were close to zero, while those on PEDOT:PSS were about 10° and those on compact TiO_2 were about 30°, with some degree of angle-hysteresis.

Yield and kinetic data, concerning hole-transfer across methylammonium lead iodide perovskite|polymer heterojunctions, have been analyzed[170] as a function of the interfacial energy-offset between the highest occupied molecular orbital of the hole-transport material and the valence band-edge of the perovskite. It was noted that just a small driving energy of about 0.07eV was needed to permit very efficient hole-transfer. Further improvement could result from using hole-transport materials having interfacial energy-offsets of between 0 and 0.18eV.

Tin-containing perovskites unfortunately tend to degrade via oxidation of the tin. They can also have very short carrier lifetimes and limited associated diffusion-lengths. Oxidation-stability, band-gaps, carrier lifetimes and solar-cell properties have been determined[171] for a wide range of compositions by varying the A-site cation and the tin/lead ratio at the B-site. The mechanisms which determine the effect of composition upon the band-gap were shown to be different to those operating in pure lead-based perovskites. In particular, alloying tin with lead greatly stabilizes the perovskite against oxidation.

Analysis of the dark current-density shows[172] that charge recombination at grain boundaries is a pivotal factor limiting the satisfactory performance of low-bandgap tin-lead halide perovskite sub-cells. Bromine incorporation can however effectively passivate the grain boundaries and decrease the dark current-density by 2 or 3 orders-of-magnitude. A suitable choice of bromine concentration can produce a tin–lead halide perovskite solar-cell with a band-gap of 1.272eV together with an open-circuit voltage deficit as low as 0.384V, a fill-factor as high as 75% and a power-conversion efficiency of better than 19%.

Materials Research Forum LLC
https://doi.org/10.21741/9781644900819

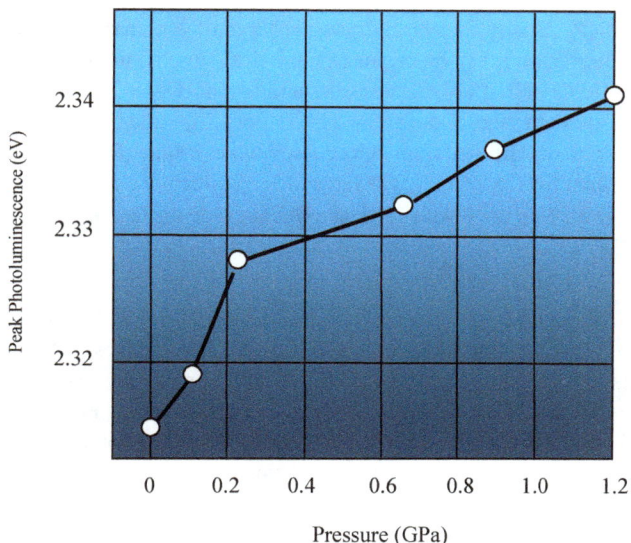

Figure 10. Energy of the main photoluminescence
peak of CH₃NH₃PbBr₃ as a function of pressure

Simple mechanical compression can affect the properties of $CH_3NH_3PbX_3$ (X = Cl, Br or I) perovskites without having to alter the composition[173]. For example, a variation in the photoluminescence as a function of pressure (figure 10) reflected an increase in the energy gap.

Density functional theory and Green's function methods have been used[174] to study the electronic and optical properties of 2-dimensional layered butylammonium lead-halide, $[CH_3(CH_2)_3NH_3]_2PbX_4$, $[CH_3(CH_2)_3NH_3]_2PbX_4$ and phenylethylammonium $(C_6H_5CH_2CH_2NH_3)$-lead halide $[C_6H_5(CH_2)_2NH_3]_2PbX_4(C_6H_5CH_2CH_2NH_3)_2PbX_4$ (X = I or Br) perovskites. Regardless of the halide in question, phenylethylammonium spacers lead to greater electronic transmission along and across lead-halide layers. Phenylethylammonium molecules are found to contribute to the density-of-states near to the band edges, and thus affect charge-carrier dynamics. Materials which contain phenylethylammonium cations exhibit less absorption in the visible range.

The use of $Cd_{1-x}Zn_xS$ as the electron-transport layer of a planar lead halide perovskite solar cell has been analyzed[175] by means of numerical simulation of a system involving a mixed-perovskite $CH_3NH_3PbI_{3-x}Cl_x$ absorber and a spiro-OMeTAD hole-transport layer, plus defects at both the $Cd_{1-x}Zn_xS|CH_3NH_3PbI_{3-x}Cl_x$ and $CH_3NH_3PbI_{3-x}Cl_x$/spiro-OMeTAD interfaces. The optimum thicknesses were 700nm for the $CH_3NH_3PbI_{3-x}Cl_x$ absorber and 50nm for $Cd_{0.2}Zn_{0.8}S$. The latter was an effective electron-transport layer for planar perovskite solar cells with a yield efficiency of up to 24.83%, an open-circuit voltage of 1.224V, a short-circuit current density of 25.283mA/cm^2 and a fill-factor of 80.22.

The distribution and charge-state of iron-induced defects in lead halide perovskite films was studied[176] using synchrotron-based X-ray techniques. X-ray absorption measurements suggested that iron-rich regions, resulting from contaminating concentrations greater than 10ppm, resembled the behaviour of Fe_2O_3. Iron within the bulk could form a mixture of Fe^{2+} and Fe^3, with the latter not being expected to be recombination-active. There was little correlation between charge collection and the presence of iron-rich regions.

Organic-inorganic perovskite films of the form, $[CH(NH_2)_2]_{0.85}(CH_3NH_3)_{0.15}Pb(Br_{0.07}I_{0.93})_3$, were prepared[177] by using a 2-step sequential method which was modified by the addition of a small fraction of dimethyl sulfoxide. By adjusting the fraction added, crystallisation and the presence of residual PbI_2 in the films could be closely controlled. Changes in dissolution due to the dimethyl sulfoxide additions altered the nucleation dynamics. Molecular intercalation occurred in the initial stages of the reaction, followed by a dissolution-crystallisation growth process. The best of the resultant devices exhibited a power-conversion efficiency of 19.29%, without hysteresis.

The power-conversion efficiencies of planar $CH(NH_2)_2PbI_3$ perovskite solar cells can be considerably improved by texturing the cell surface with columnar hollow nano-arrays, conical hollow nano-arrays, square prism hollow nano-arrays and pyramidal hollow nano-arrays. Upon comparing the power-conversion efficiencies of planar cells having the same layer-depth, it is found[178] that when the layer depth is between 50 and 125nm and when the array period and areal fraction of the nanotexture are optimum, the efficiency increases from 29 to 50% for the columnar hollow and square prism hollow textured cells, relative to planar ones. The increase was from 20 to 41% for the conical hollow and pyramidal hollow textured cells relative to planar ones. When the layer-depth was less than 25nm, the ultimate efficiencies of the columnar hollow and square prism hollow textured cells were greater than those of conical hollow and pyramidal hollow ones.

A theoretical study has been made[179] of point-defect screening by $CH_3NH_3^+$ dipoles in $CH_3NH_3PbX_3$ (X = I, Br, Cl) perovskites, on the basis of a statistical model and short-range and dipolar interactions between the $CH_3NH_3^+$ cations. The model was extended so as to account for long-range charge-dipole interactions between defects and organic cations. Charge-screening in various structural phases of $CH_3NH_3PbI_3$ perovskite was studied for various values of the charge-dipole interaction energy. An appreciable interaction disturbed the antipolar long-range order of the $CH_3NH_3^+$ cations; thus giving rise to a multidomain phase having a small electric polarization.

Organic and inorganic materials are frequently combined in hybrid devices, and the identification of layers and interface defects by depth-profiling can be difficult, especially in the case of perovskite solar cells with their metal electrodes, mesoscopic conductive oxide, already-hybrid absorber and organic hole-extraction layer. The profiles obtained using monatomic Ar^+ beams of differing energies have been compared[180] with those obtained using argon-ion clusters having between 150 and 1000 members and energies of up to 8keV. A systematic study was made of $CH_3NH_3PbI_3$ solar cells and $\{[CH(NH_2)_2]_xCs_{1-x}PbI_3\}_{0.85}(CH_3NH_3PbBr_3)_{0.15}|TiO_2$. In the case of monatomic beams, the implantation of positively charged atoms provoked the surface diffusion of free iodine species from the perovskite, thus changing the iodine/lead ratio. Lead atoms which were in the metallic Pb^0 state accumulated at the bottom of the perovskite layer, where the Pb^0/Pb_{tot} fraction attained 50%. In the case of argon clusters, ion-beam induced iodine diffusion was reduced only when the etch-rate was high enough to guarantee a profile duration which was comparable to that of low-energy Ar^+ ions. Suitable erosion-rates were found only for 300 and 500 membered clusters at 8keV, but Pb^0 particles in the perovskite were then sputtered less efficiently, leading to an increase in the Pb^0/Pb_{tot} fraction of up to 75% at the perovskite|TiO_2 interface.

Photo-electron spectroscopy and visible laser illumination have been used[181] to study the photo-stability of perovskite films having Br:I ratios of 50:50 and 17:83, and compositions with and without Cs^+. In samples without cesium, and in 50:50 samples, the surface was enriched in bromine and depleted in iodine during illumination. Some of the perovskite decomposed to give Pb^0, organic halide salts, and iodine. Both reactions were partially reversible following illumination.

The carrier diffusion length in $CH_3NH_3PbI_3$ and $CH_3NH_3PbI_{3-x}Cl_x$ perovskites was measured[182] by using scanning photocurrent microscopy and fitting the data to an exponential curve. The diffusion length of the mixed-halide perovskite was higher than that of the iodide perovskite film by a factor of 3 to 6; consistent with the assumption that the carrier mobility would be higher in mixed lead-halide perovskites. The diffusion length could be explained in terms of a drift-diffusion model.

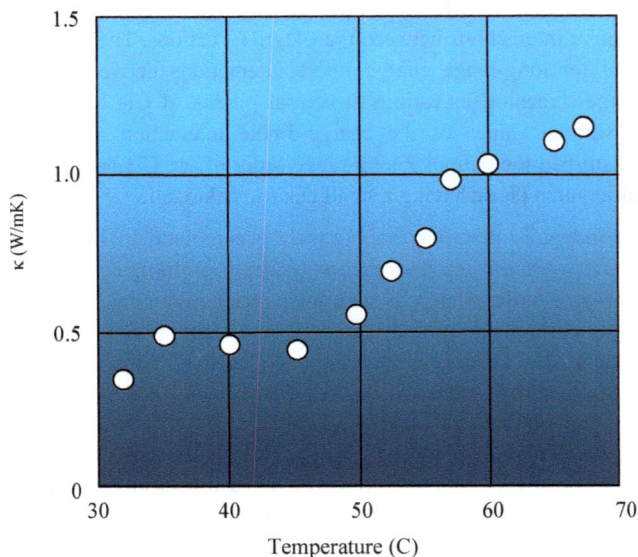

Figure 11. Thermal conductivity of cleaved $CH_3NH_3PbI_3$ single crystal

The identification of crystalline phases by using *in situ* X-ray diffraction during growth was used to monitor[183] phase evolution during the heat-treatment of $CH_3NH_3PbX_3$ (X = I, Br or Cl) perovskite thin films. The latter were prepared by vacuum-based 2-source co-evaporation of various methylammonium halide and lead halide precursors. Single-halide perovskites $CH_3NH_3PbI_3$, $CH_3NH_3PbBr_3$ and $CH_3NH_3PbCl_3$ were prepared without any secondary phases. At a substrate temperature of 120C, the halides in $CH_3NH_3PbI_3$/$CH_3NH_3PbBr_3$ thin films could be completely and reversibly exchanged during exposure to each other.

The thermal conductivities of methylammonium lead halide perovskite, $CH_3NH_3PbX_3$ (X = I, Br or Cl), single crystals and thin films were determined[184] by means of scanning near-field thermal microscopy. The room-temperature thermal conductivity of $CH_3NH_3PbX_3$ single crystals was found to be 0.34, 0.44 and 0.50W/mK for X = I, Br and Cl, respectively. Surprisingly similar thermal conductivities (figure 11) were found for the corresponding thin films. The thermal conductivity of $CH_3NH_3PbI_3$ in the cubic phase

Materials Research Forum LLC
https://doi.org/10.21741/9781644900819

at above 55C increased to 1.1W/mK. The absolute value of the negative linear thermal expansion coefficient (figure 12) of the tetragonal structure was -5.1 x 10^{-4}/K, and was quite similar to the positive linear thermal expansion coefficient of the cubic structure: 4.6 x 10^{-4}/K.

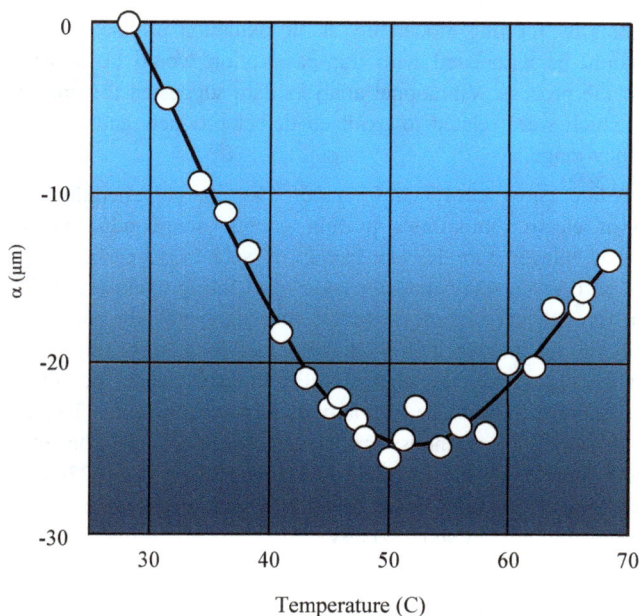

Figure 12. Linear thermal expansion coefficient of $CH_3NH_3PbI_3$

The self-catalyzed vapour-liquid-solid growth of (PbX$_2$, X = Cl, Br or I) nanowires, and their conversion to perovskites, have been described[185] in terms of a kinetic model in which a liquid lead catalyst was supersaturated with a halogen via the vapour-phase introduction of both lead and the halogen; thus triggering the growth of nanowires. In the case of PbI$_2$, the nanowires were monocrystalline and were oriented in the [12•0] direction. They consisted of a stoichiometric PbI$_2$ shaft, with a spherical lead tip. Low-temperature vapour-phase intercalation of methylammonium iodide then converted the nanowires into the $CH_3NH_3PbI_3$ perovskite while maintaining the nanowire morphology. The nanowires exhibited a large optical antenna effect, thus leading to markedly increased scattering and absorption efficiencies. These could be more than twice those of thin films having the same thickness.

A large increase in the photoluminescence quantum-yield was reported[186] to occur upon exposing these perovskites to molecular oxygen. Based upon density functional theory, it has been proposed that oxygen can inactivate deep hole-traps which are associated with iodide interstitials by forming moderately stable oxidation products. The small energy gain which would be associated with trap-passivation would be in accord with the reversibility of the process. Vibrational analysis data suggested the appearance of new active modes which were related to oxidized defect products and covered the 300 to 700/cm frequency-range.

Hybrid lead halide perovskites can be used[187] as charge-generation layers for the determination of electron mobilities in thin organic semiconductors. Time-of-flight measurements of bilayer samples consisting of such a layer of charge-generating perovskite and an organic semiconductor layer of differing thickness are determined by carrier motion through the organic material; consistent with the much higher charge-carrier mobility in the perovskite. When combined with efficient photon-to-electron conversion in the perovskite, the high mobility-imbalance then permits electron-only mobility measurements to be made of relatively thin organic films. This is not possible using other time-of-flight methods. As a demonstration, electron mobilities were determined as a function of electric field and temperature, in a 127nm-thick layer of electron-transporting perylene di-imide based polymer, which were consistent with an exponential trap distribution of about 60meV.

A wavelength- and illumination-intensity-dependence of the hysteresis in cells with an 18% power-conversion efficiency has been demonstrated[188]. The perovskite devices exhibit a lower hysteresis under illumination with near band-edge (red) wavelengths as compared with more energetic (blue) excitation. This was explained in terms of thermalization-assisted ion migration or vacancy generation, as supported by impedance-spectroscopy data and the dependence of the photovoltage decay upon the illumination-time and excitation-wavelength. It was suggested that high-energy photons create hot charge-carriers that can create additional vacancies by thermalization, or release more energetic phonons that help to surmount the activation energy for ion migration.

Non-radiative recombination pathways can play an important role, and appear as photoluminescence-inactive areas on perovskite films. Wide-field photoluminescence microscopy and impedance spectroscopy of perovskite films have been used[189] to investigate the spatial and temporal evolution of such photoluminescence-inactive areas, when under the influence of an external electric field. The formation of the photoluminescence-inactive domains was subsequently attributed to the migration and accumulation of iodide ions under the external field. It was noted that iodine vapour directly affects photoluminescence-quenching of a perovskite film, thus indicating that the migration and segregation of iodide ions play important roles in the photoluminescence-quenching and limits the power-conversion efficiency of organometallic halide-based perovskite photovoltaic devices.

In simulations of a $CH_3NH_3PbI_3$ perovskite-based single-junction solar cell, it was observed[190] that there was a systematic variation in device performance due to variations in the optical absorption of the active layer. Upon increasing the thickness of the absorber layer from 50 to 1000nm, the power-conversion efficiency changed from 7.9 to 21.1%, the open-circuit voltage changed from 1.26 to 1.16V and the short-circuit current-density changed from 7.56 to 22.61mA/cm^2 while the fill-factor remained constant at 83%.

It has been found that benzaldehyde can preserve the stoichiometry of precursors in solution during perovskite preparation. This leads to improvements in perovskite solid-state film morphology, and in the performance of $CH_3NH_3PbI_{3-x}Cl_x$-based solar cells[191].

Perovskite photovoltaics have been prepared[192] by partially replacing the lead cation with calcium ions. The band-gap narrowed with increasing additions of calcium. Perovskite film which was doped with 1.0mol% of calcium exhibited the lowest emission energy and recombination behavior. The open-circuit voltage also increased from 0.93 to 0.98V, the short-circuit current density increased from 17.4 to 19.1mA/cm^2 and the power-conversion efficiency increased from 10.7 to 12.9%.

The power-conversion efficiency, hysteresis and device lifetime of $CH_3NH_3PbI_{3-x}Cl_x$ solar cells can all be improved[193] by incorporating polymethyl methacrylate into the electron-extraction layer. By using polymethyl methacrylate having a suitable molecular weight, a 30% improvement of the power-conversion efficiency could be obtained together with decreased hysteresis and device degradation. The improved properties were attributed to surface passivation of the $CH_3NH_3PbI_{3-x}Cl_x$, leading to lower charge-trapping at the cathode interface and Shockley-Read-Hall charge-recombination.

The $CH_3NH_3X_3$ perovskites have very long photogenerated charge-carrier diffusion lengths, and a low recombination rate. Frequency- and temperature-dependent dielectric measurements performed[194] over the entire frequency spectrum have shown that the dielectric constant takes values of more than 27 at frequencies below 1THz, regardless of the halide involved. This prevents photocarrier-trapping and recombination, due to strong screening of the charged components. Large contributions to the dielectric constant were attributed to dipolar disorder of the $CH_3NH_3^+$ cations and lattice dynamics in the GHz range. This led to room-temperature dielectric constants of 62 for the iodide, 58 for the bromide and more than 45 for the chloride below 1GHz. The disorder decreased continuously with decreasing temperature.

Organic-inorganic hybrid perovskites tend to have complex compositions. A 4-state model of the photoluminescence yield has been developed[195] in which optical transitions and emission spectra are deconvoluted in order to quantify the band-gaps and charge-quenching yields in such materials. The model has been applied to a charge-quenching yield analysis and correlated with the solar-cell performances of $CH_3NH_3PbI_3$ and $CH_3NH_3PbI_{3-x}Cl_x$ with mesoporous TiO_2 layers. The inclusion of chlorine here improves crystal formation.

A theoretical approach, based upon drift-diffusion equations, has been used[196] to study planar mixed lead halide perovskite solar cells. A bulk defect density of about $10^{16}/cm^3$, and surface recombination velocities of the order of 10cm/s, were deduced and this agreed well with experimental data. It was concluded that the overall efficiency of perovskite solar cells is governed mainly by the open-circuit voltage, and that interface defects constitute an important loss factor.

Most of the organic conductors, such as spiro-OMeTAD and PTAA, which are incorporated into high-efficiency perovskite solar cells have a low thermal stability. Copper phthalocyanine has been considered to be a good hole-transport material because it exhibits excellent thermal stability and good interfacial bonding. Solar cells which incorporate copper phthalocyanine have exhibited a power-conversion efficiency of about 18% and have retained 97% of their initial efficiency for more than 1000h during annealing at 85C. The cells were also stable under thermal cycling for 50 times at between -45 and 85C. The high power-conversion efficiency and thermal stability were due to a strong interfacial coating on the surface of perovskite facets located between the copper phthalocyanine and the perovskite layer[197].

The effect of low-energy, 10 to 20keV, electron bombardment upon lead halide perovskites is to cause radiolysis[198]. Discontinuous metallic lead layers form at the top and bottom surfaces of perovskite microplates during radiolysis, and can act as carrier-conducting layers; thus increasing the photocurrent in a perovskite photodetector by 217%.

The effect of $PbCl_2$, in a PbI_2 solution of N,N-dimethylformamide, upon the properties of thin films was investigated[199], showing that a planar perovskite solar cell based upon 300nm-thick $CH_3NH_3I:PbI_2:PbCl_2$ (4M:1.25M:0.75M) thin film exhibited a photo-electric conversion efficiency of 6.3% together with an open-circuit voltage of 0.66V, a short-circuit photocurrent density of 19.60mA/cm^2 and a fill factor of 58.3%.

A detailed experimental study has been made[200] of hole-injection from $CH_3NH_3PbI_3$ to various hole-transport layer materials. Those carrier dynamics which were directly related to hole-injection were identified by using a pump light having a short absorption depth, and comparing the transient transmission signals which were excited on each side of the sample. This showed that hole-injection into PTAA and PEDOT:PSS was completed within 1 and 2ps, respectively. Hole-injection into NiO_x involved an additional slow process having a 40ps time-scale.

Density functional theory calculations were performed[201] for all of the possible ion-migration barriers in perovskites involving various cations: $CH_3NH_3PbI_3$, $CH(NH_2)_2PbI_3$ and $CsPbI_3$. The most relevant ionic motion, that of the iodide, was treated in more detail. There was a correlation between the energy barrier to iodine migration, and the size of the dipole of the monovalent cation. A vacancy-dipole interaction mechanism was proposed in which the larger dipole of the monovalent cation could respond to, and screen, the local electric fields more effectively. The stronger response of a highly dipolar monovalent cation to the vacancy electrostatic potential could then lead to lesser structural changes within neighbouring octahedra.

It has been noted that comparable efficiency losses occur in lead halide perovskites at iron-contents which are some 100 times greater than those in p-type silicon. Photoluminescence measurements link iron concentrations to a non-radiative recombination which is attributed to the presence of deep-level iron interstitials and iron-rich particles[202]. At moderate contamination levels, there was a clear recovery of device efficiency upon biasing at 1.4V for 60s in the dark. It was suggested that this temporary effect arose from charge-carrier collection which was improved by electric fields that were strengthened due to ion migration toward the interfaces.

Efficient mixed perovskite solar cells which were based upon $(CH_3NH_3)_{0.7}[CH(NH_2)_2]_{0.3}Pb(I_{0.9}Br_{0.1})_3$ were produced[203] by optimizing the annealing conditions. A better film quality was obtained by annealing at 100C for 0.5h. The corresponding device exhibited a power-conversion efficiency of 16.76%, an open-circuit voltage of 1.02V, a short-circuit current density of 21.55mA/cm^2 and a fill-factor of 76.27%. The $(CH_3NH_3)_{0.7}[CH(NH_2)_2]_{0.3}Pb(I_{0.9}Br_{0.1})_3$ exhibited a greatly increased thermal stability. Following heating at 80C for 24h, its power-conversion efficiency retained 70% of its initial value. In the case of the control-sample, $CH_3NH_3PbI_3$, only 46.50% of the initial value could be retained.

Photo-excited dynamic processes in $CH_3NH_3PbI_3$ quantum-dots have been modelled[204] by using many-body perturbation theory and non-adiabatic dynamics. The photo-excitation was assumed to be followed by exciton-cooling, radiative or non-radiative recombination or multi-exciton generation. The time-scales which were calculated for those processes could be arranged in the order: multi-exciton generation < exciton-cooling < radiative < non-radiative recombination. The estimate for the first of these was of the order of some femtoseconds, while exciton-cooling was in the picosecond range and the others were in the nanosecond range. These figures showed which electronic transition pathways could contribute to an increase in charge-collection efficiency. It was concluded that quantum-confinement promotes multi-exciton generation in $CH_3NH_3PbI_3$ quantum dots.

Two-dimensional layered hybrid perovskites have a higher exciton binding energy and a possibly higher light-emission efficiency than do 3-dimensional perovskites. The dissolution–recrystallization growth of 2-dimensional perovskites, based upon 2-phenylethylammonium ($C_6H_5CH_2CH_2NH_3^+$) cations, led to monocrystalline microplates having a well-defined rectangular geometry of nanoscale thickness[205]. The crystal structures of the $(C_6H_5CH_2CH_2NH_3)_2PbX_4$ (X = Br or I) perovskites were confirmed using monocrystal X-ray diffraction. By using mixed-halide compositions, the photoluminescence emission of $(C_6H_5CH_2CH_2NH_3)_2Pb(Br,I)_4$ perovskites with a narrow peak bandwidth could be varied from 410 to 530nm.

The as-prepared perovskites, hydroxyl ammonium lead iodo-chloride and hydroxyl ammonium lead chloride, have band-gaps of 3.7 and 3.9eV, respectively. Time-resolved photoluminescence data show[206] that $OHNH_3PbCl_3$ exhibits long-lived photoluminescence whereas $OHNH_3PbI_2Cl$ has a short photoluminescence lifetime. They are both stable, and show no signs of decomposition at up to about 200C. Photocatalytic data, gathered in sunlight and monitored using direct yellow textile dye, showed that $OHNH_3PbI_2Cl$ degraded the dye in 20min and $OHNH_3PbCl_3$ in 55min.

A study of as-deposited or annealed $CH_3NH_3PbBr_3$ and $CH_3NH_3PbI_3$ perovskite films, processed by interdiffusion, showed[207] that the absorption spectra exhibited broad band edge saturation in the case of as-deposited films. Sharp excitonic absorption was observed in the case of annealed films. The fluorescence emissions of the films exhibited a marked dependence upon the type of halogen involved, with a quantum-yield of about 90% for annealed $CH_3NH_3PbBr_3$ film, and slightly Stokes-shifted bands. The former behaviour was attributed to its crystallinity and to the quantum confinement of excitons.

Density functional theory calculations have been made[208] of $CH_3NH_3PbI_{3-x}Cl_x$ layers, deposited onto tetragonal $PbTiO_3$ (001) surfaces, showing that ferroelectric polarization - pointing from the $PbTiO_3|$perovskite interface to $PbTiO_3$ – favoured photogenerated electron-transfer across the interface, and their transport to the collecting electrode. The $PbTiO_3$ internal electric field meanwhile led to a position-dependent energy-level diagram. The perovskite gap could be changed by chlorine concentrations at the interface and by the surface terminations of $PbTiO_3$ and perovskite layers.

The carrier transport in $CH_3NH_3PbI_3$-based organic–inorganic perovskites is complicated by vacancy-mediated ion migration. Temperature-dependent pulsed voltage versus current measurements of thin films, performed under dark conditions, have been used[209] to decouple ion-migration, ion-accumulation and electronic transport. The electric-field history and scan rate affected the electronic transport via a mechanism which involved ion-migration and accumulation at electrode interfaces. Thermally-activated processes, having activation energies of 0.1 and 0.41eV, were detected and were attributed to the electromigration of iodine and methylammonium vacancies, respectively. Electromigration of these ionic species was deemed to be responsible for modifying the interfacial electronic properties of $CH_3NH_3PbI_3$. It was concluded that the intrinsic behavior of $CH_3NH_3PbI_3$ was responsible for the hysteresis of solar cells,

A combination of experimental measurements and theory has shown[210] that the evaporation-rate of dimethylformamide solvent during the spin-coating of a mixed lead halide precursor is 1.2×10^{-8} m/s. When K-bar coating the same precursor, the solvent does not evaporate appreciably during deposition and a resultant rough film leads to a power-conversion efficiency of less than 2%. Upon increasing the air flow, of a K-bar coated perovskite film during crystallisation to 2.7×10^{-4} m/s, the power-conversion efficiency increases to 8.5% due to improvements in the short-circuit current and fill-factor.

It has been shown[211] that the Mayer-bond order of the polar atoms of a solvent is a reliable guide to predicting the solubility of lead halide perovskite precursors in the solvents used for their preparation, and that it is much more reliable than the Hansen polar solubility parameter.

Low trap-density monocrystalline $CH_3NH_3PbBr_3$ perovskite, prepared via inverse temperature crystallization, has been used[212] to avoid the interfering effects of defects. It was shown that inductive behavior predominates in the frequency range where the capacitance takes negative values. Under high bias, there were shown to be 2 bias-dependent inductive elements, related to Br^- and $CH_3NH_3^+$ ions, participating in vacancy-assisted ionic diffusion in perovskite crystals.

Nanoscale spatial distributions of PbI_2 on $[CH(NH_2)_2PbI_3]_{0.85}(CH_3NH_3PbBr_3)_{0.15}$ perovskite thin films have been investigated[213] by using a method which permitted the simultaneous mapping of the morphologies of perovskite and PbI_2 grains by selectively detecting their characteristic fluorescence bands using band-pass filters. Perovskite samples, with and without excess PbI_2, revealed a PbI_2 distribution in the case of PbI_2-

rich samples. Nanoscale time-resolved photoluminescence techniques showed that the PbI_2-rich regions had a longer lifetime, due to suppressed defect-trapping, as compared with PbI_2-poor regions. This showed that the passivating-effect of PbI_2 in perovskite films is especially effective in localized regions.

A study of the perovskite crystallization-precursors, $PbCl_2$, $PbBr_2$ and PbI_2, has shown[214] that the volatility which is important for thin-film growth, increases with the size of the halogen ion (table 6). Deposited thin films of PbI_2 exhibited good optical properties and had a band-gap energy of 2.3eV.

Table 6. Vapour Pressures of lead halide perovskite precursors

Halide	Temperature (C)	Vapour Pressure (Pa)
$PbCl_2$	655.2	0.128
$PbCl_2$	663.6	0.195
$PbCl_2$	668.6	0.249
$PbCl_2$	673.6	0.312
$PbCl_2$	678.6	0.390
$PbCl_2$	683.6	0.501
$PbCl_2$	688.6	0.631
$PbCl_2$	693.6	0.780
$PbCl_2$	698.6	0.956
$PbCl_2$	703.6	1.203
$PbBr_2$	603.2	0.128
$PbBr_2$	606.1	0.15
$PbBr_2$	608.6	0.173
$PbBr_2$	611.1	0.197
$PbBr_2$	613.6	0.226
$PbBr_2$	616.1	0.256
$PbBr_2$	618.6	0.294

$PbBr_2$	621.1	0.331
$PbBr_2$	623.6	0.379
$PbBr_2$	626.1	0.433
$PbBr_2$	628.6	0.491
$PbBr_2$	631.1	0.555
$PbBr_2$	633.6	0.632
PbI_2	579.7	0.0920
PbI_2	588.2	0.149
PbI_2	591.1	0.174
PbI_2	595.2	0.220
PbI_2	598.3	0.260
PbI_2	602.1	0.328
PbI_2	607.5	0.435
PbI_2	612.2	0.568
PbI_2	617.3	0.734
PbI_2	622.3	0.960
PbI_2	626.1	1.148

When nanoporous TiO_2-based lead iodide perovskite solar cells were exposed[215] to a range of light intensities, the short-circuit photocurrent increased linearly with intensity while the fill-factor decreased slightly. A logarithmic plot of the light-intensity dependence of the open-circuit voltage had a slope of 1.09kT/q. It was suggested that the solar-cell performance could be improved simply by increasing the incident-light intensity.

Noting that methylammonium lead halide perovskites were usually deposited from low-volatility high boiling-point solvents such as N,N-dimethylformamide, where slow drying led to the formation of large crystallites which increased surface roughness and the incidence of pin-holes, it was pointed out[216] that the more-volatile 2-methoxyethanol would produce smaller crystals. This would improve the surface coverage of perovskite films, reduce leakage currents and increase the open-circuit voltage and fill-factor of solar cells. Cells with a p-i-n configuration, made using a triple-anion precursor salt, led to an increase in the power-conversion efficiency from 14.1 to 15.3% when N,N-dimethylformamide was replaced by 2-methoxyethanol.

A study of the effect of $TiCl_4$ aqueous solutions on the performance of $CH_3NH_3PbI_{3-x}Cl_x$ perovskite solar cells showed[217] that all of the cell parameters increased following treatment with $TiCl_4$ up to a concentration of 100mM. An improvement in the fill-factor was attributed mainly to a decrease in the series resistance of the treated cells. More efficient transport and suppressed charge-recombination in the treated cells occurred, leading to an increase in the short-circuit current density. A maximum efficiency of 15.6% was observed for a cell which was post-treated with 100mM $TiCl_4$ solution; a 25% improvement over that of untreated cells.

Part of the cause of the large hysteresis observed in the current-voltage scans of methylammonium lead iodide perovskite solar cells has been attributed[218] to screening of the built-in field by mobile ions. Photocurrent transients, measured within 100μs of a voltage-step, have furnished direct evidence that field-screening exists. Just after a step to forward-bias, the photocurrent transients reversed in sign and these inverted transients could place an upper bound on the width of the space-charge layers next to electrodes. The lower bound on the mobile-charge concentration was deduced to be greater than 1 x $10^{17}/cm^3$. As to be expected of mobile ions in a solid electrolyte, the space-charge layer-thickness remained approximately constant as a function of bias.

An investigation of controlled PbI_2 formation in $CH_3NH_3PbI_3$ photovoltaics demonstrated[219] the beneficial role played by PbI_2 in device performance. High-resolution transmission electron microscopy revealed the location of PbI_2 in the active layer. During annealing, PbI_2 formed mainly in the grain-boundary regions of films. At certain temperatures, the PbI_2 which formed could be very beneficial by reducing current-voltage hysteresis and increasing the power-conversion efficiency.

A one-step slot-die coating method is suitable for the deposition of lead halide perovskite layers, and especially for infiltration into a mesoporous titania scaffold. The crystallisation of the perovskite can be controlled[220] via the substrate temperature, and devices processed in this way have power-conversion efficiencies of up to 9.2%.

An alcohol-soluble conjugated bispyridinium bromide can be used as a cathode-modifier in order to improve the cathode interface of planar heterojunction perovskite solar cells[221]. The notable electron-extraction ability of bispyridinium rings gives bispyridinium bromide a favorable energy-level alignment with phenyl-C60-butyric acid methyl ester and a cathode such as aluminium. This produces an ideal ohmic contact plus efficient electron transport and collection. The deep-lying highest occupied molecular orbital energy level of bispyridinium bromide can also block hole carriers and thereby decrease leakage currents and hole-electron recombination at the cathode interface. Bispyridinium bromide can n-dope C60-butyric acid methyl ester via anion-induced electron transfer and sharply increase the electron mobility of C60-butyric acid methyl ester. This decreases the interfacial resistance and promotes electron transport. By incorporating a bispyridinium bromide cathode interlayer, low-hysteresis solar cells having a maximum power-conversion efficiency of 19.61% can be produced. Devices which omit this cathode interlayer exhibit a power-conversion efficiency of 16.97%.

Band-edge carrier dynamics in single crystals of $CH_3NH_3PbBr_3$, $CH(NH_2)_2PbBr_3$ and $CsPbBr_3$ have been investigated[222] by using time-resolved photoluminescence and transient reflectance spectroscopic methods. The results indicated similar low carrier-trapping rate-constants and electron–hole radiative recombination rate-constants for all of the crystals, regardless of the cation-type. Measurements of the surface recombination velocity and carrier mobility in the near-surface region of those single crystals again yielded similar results for all of them.

Theoretical and experimental studies have been made[223] of the effect of substituting transition metals at the lead site of $CH_3NH_3PbX_3$ (X = Br or Cl). Computations identified iron- and cobalt-substituted $CH_3NH_3PbBr_3$ as being suitable absorbers, with a mid-gap density-of-states. Steady-state and time-resolved photoluminescence studies revealed no sign of self-quenching for cobalt substitutions of up to 25%.

In order to examine the dependence of diffusion length on carrier density, direct and independent measurements were made[224] of the carrier-diffusion coefficient and recombination-rate in methylammonium lead halide perovskite layers, using the light-induced transient grating technique. This revealed the existence of 2 distinct carrier-diffusion regimes for densities ranging from 10^{18} to $10^{20}/cm^3$. In perovskite films of high compositional quality, diffusion was governed by the band-like transport of free carriers. The diffusivity was 0.28 to $0.7cm^2/s$, even at low carrier densities, and increased with excitation due to carrier degeneracy. In disordered layers, the diffusion was governed by hopping-like transport of localized carriers. The diffusion coefficient in those layers was 0.01 to $0.04cm^2/s$ at low densities and increased with excitation due to local state filling and carrier delocalization. The carrier recombination could be described in terms of non-linear radiative and non-radiative recombination coefficients which saturated, with excitation, due to phase-space filling at high carrier densities.

The high power-conversion efficiencies of hybrid $CH_3NH_3PbX_3$ (X = I, Br or Cl) solar cells are thought to be closely related to the behaviour of the methylammonium cations. A statistical phase-transition model could accurately describe[225] ordering of the $CH_3NH_3^+$

cations and the phase-transition sequence of the $CH_3NH_3PbI_3$ perovskite. The model posited the existence of short-range strain-mediated and long-range dipolar interactions between the cations. Monte Carlo simulations performed on a 3-dimensional lattice predicted the heat capacity and electric polarization of $CH_3NH_3^+$ cations. The temperature-dependence of the polarization revealed the antiferroelectric nature of the perovskites.

The morphology of methylammonium lead halide perovskite thin films markedly affects the performance of related devices. The presence of excess methylammonium iodide during perovskite crystallization leads to the formation of layered intermediates which are related to the formation of large continuous grains[226]. When the precursors are present in stoichiometric ratios, the initial stages of crystallization involve the formation of solvates. The crystallized films then exhibit poor surface coverage and contain needle-like structures. The activation energy of crystallization in films with excess methylammonium iodide is higher than that for films with stoichiometric precursor ratios. The higher activation energy is consistent with having to sublime the excess methylammonium iodide during film formation. The replacement of iodide with chloride does not affect the morphology, although it decreases the activation energy for crystallization. This could be attributed to the lower sublimation energy of methylammonium chloride.

Under open-circuit conditions, at least one electron-hole pair to photon to electron-hole pair recycling event occurs in these solar cells during solar irradiation[227]. This can lead to a considerable reduction in the external photoluminescence yield as compared with the internal yield. It has been shown that, given an internal yield of 70%, the external yield can be as low as 15% in planar films where light out-coupling is inefficient. Values which were as high as 57% were found for films having textured substrates. A detailed analysis of photo-excited carrier dynamics, radiative and non-radiative recombination events and photoluminescence efficiencies led to the conclusion that the use of textured active layers improve power-conversion efficiencies.

A study was made[228] of the morphology and structural properties of spin-coated $CH_3NH_3PbI_3$ perovskite films which had been spin-coated onto a Si/SiO$_2$|poly(3,4-ethylenedioxythiophene)polystyrene sulfonate (PEDOT:PSS) substrate and held at various temperatures. Single perovskite grains, buried within a complex polycrystalline film, could be investigated. This showed that spin-coating perovskite precursor solutions at higher temperatures led to an improved surface coverage and a better solar-cell performance.

It has been clearly demonstrated[229] that perovskite layers remain unaltered during device use. On the other hand, considerable changes in the absorption spectra of the spiro-OMeTAD layer which is routinely used as the hole-transport layer can gradually cause a loss of performance, due to its degradation. Photo-induced oxidation can occur in both air and in an inert atmosphere and this is accelerated by the chemicals which are routinely added to the spiro-OMeTAD and by intimate contact with the electron-injecting TiO_2 layer.

Study of the luminescence of polycrystalline thin films of the perovskites shows[230] that, in an inert environment, photo-induced formation of emissive sub-bandgap defect states occurs, regardless of the composition of the lead halide semiconductor, and quenches the band-to-band radiative emission. Carrier-trapping takes place on a sub-ns time-scale, and trapped carriers recombine within a few μs. The presence of even a very small amount of oxygen is able to compensate for such an effect.

Pure chlorobenzene and 2-propanol spin-coating have been used (possibly mixed) in the single-step preparation of $CH_3NH_3PbI_3$ and $CH_3NH_3PbI_{3-x}Cl_x$ films, respectively[231]. Planar heterojunction solar cells which were based upon $CH_3NH_3PbI_3$ prepared using a mixed-solvent technique offered a poor power-conversion efficiency due to the formation of small numbers of perovskite crystals which suffered moreover from low crystallinity. The use of pure chlorobenzene imparted the corresponding devices with an efficiency of 8.1%, due to the formation of uniform $CH_3NH_3PbI_3$ thin films having higher surface coverages and superior crystallinity. Cells which were based upon $CH_3NH_3PbI_{3-x}Cl_x$ prepared using mixed solvents led to the highest power-conversion efficiency of 9.2%, a short-circuit current-density of $16.06 mA/cm^2$ and a fill-factor of 63.6%. There was little photocurrent hysteresis, due to better light absorption. This in turn was attributed to the larger crystal size, better surface morphology and absence of pin-holes.

A study of the effect of temperature upon the properties of $CH_3NH_3PbI_3$, $CH_3NH_3PbBr_3$ and $CH(NH_2)_2PbBr_3$ showed[232] that the photoluminescence of $CH_3NH_3PbI_3$ and $CH_3NH_3PbBr_3$ at temperatures below 100K exhibited 2 distinct emission peaks. The $CH(NH_2)_2PbBr_3$ instead featured a single emission peak. Regardless of the composition, the band-gap exhibited an unexpected blue-shift upon increasing the temperature from 15 to 300K. The additional photoluminescence peak was attributed to the presence of molecularly disordered orthorhombic domains, and the unexpected blue-shift of the band-gap with increasing temperature was attributed to stabilization of the valence-band maximum.

Two-dimensional chemical-mapping methods revealed[233] that perovskite grain boundaries consisted of non-stoichiometric PbI_x or $CH_3NH_3PbI_x$ which exhibited an absence of chloride, an increased oxygen content and a high density of iodide vacancies. Steady-state 2-dimensional photoluminescence data revealed the occurrence of band-gap broadening at the grain boundaries, and 2-dimensional lifetime-mapping suggested that the boundaries contained deep defect-centers. These defective grain boundaries did not act as recombination sites for photogenerated charge-carriers due to band-gap broadening of non-stoichiometric PbI_x or $CH_3NH_3PbI_x$ at the grain boundary. Photo-generated charge-carriers adjacent to the boundaries were expected to be repelled by those boundaries and result in a greater reduction in the recombination of photogenerated charge carriers. This was suggested to be a possible explanation for the good performance of $CH_3NH_3PbI_{3-x}Cl_x$-based solar cells.

Kelvin-probe force-microscopic studies of the effects of heat-treatment upon $CH_3NH_3PbI_3$ revealed[234] the existence of a positive potential barrier – apparently due to chemical inhomogeneity – at the grain boundaries; a factor which is known to have the beneficial effect of suppressing carrier recombination. The height of the barrier here increased with increasing annealing time, and this could be directly related to device performance. A similar temperature-dependence of the current-density as a function of biasing occurred, regardless of annealing time. All efficiency then effectively disappeared at low temperatures due to a divergent series resistance. The latter was tentatively attributed to a freeze-out of free carriers or to the presence of Schottky-type barriers at the interface.

It was noted[235] that the coordination of lead to competing solvent molecules and iodide ions governed the type of complexes which were present in films, and $PbIS_5^+$, PbI_2S_4, $PbI_3S_3^-$, $PbI_4S_2^{2-}$, $PbI_5S_2^{3-}$, PbI_6^{4-} and 1-dimensional $(Pb_2I_4)_n$ chains could all be detected. It was concluded that strongly-coordinating solvents preferentially formed species which involved a low number of iodide ions, while more weakly coordinating solvents generated high concentrations of PbI_6^-, with all such ions acting as structural defects which could determine the electronic properties of photovoltaic films.

In early work, organic lead halide perovskite solar cells were prepared which included a transparent sputtered indium tin oxide upper electrode[236]. A power-conversion efficiency of 1.5% was measured for transparent cells with a thin molybdenum oxide buffer layer and indium tin oxide electrode. An efficiency of 2% was found even in cells with the tin oxide electrode sputtered directly onto the organic charge-transport layer.

Knudsen effusion-mass spectrometric studies of $CH_3NH_3PbCl_3$, $CH_3NH_3PbBr_3$ and $CH_3NH_3PbI_3$ showed[237] that they all decomposed into the corresponding solid lead halide plus the gaseous hydrogen halide and methylamine; a decomposition which began at temperatures as low as 60C. This was then seen as a potential problem for the long-term use of solar devices.

Carrier dynamics in $CH_3NH_3PbI_{3-x}Cl_x$ thin films having various crystalline morphologies were studied[238] as a function of temperature and excitation wavelength. Room-temperature long-lived (> 1ns) transient absorption signals indicated the occurrence of negligible carrier-trapping. Ultrafast 400nm-excited photoluminescence data meanwhile revealed a large-amplitude 45ps decay which persisted down to the orthorhombic phase-transition temperature. The rapid band-edge photoluminescence decay appeared only for excitations greater than 520nm. Lower photon-energy excitation led to slow dynamics which were consistent with negligible carrier-trapping. The results suggested that the rapid photoluminescence decay arose from the excitation of high-energy phonon-modes which were associated with the organic sub-lattice. The rapid photoluminescence decay was attributed to organic-to-inorganic sub-lattice equilibration.

The controllable formation of grain-boundary PbI_2 nanoplate-passivated $CH_3NH_3PbI_3$ and $CH_3NH_3PbI_2Br$ perovskites was shown[239] to be possible, yielding solar cells having efficiencies of up to 17.8 and 14.4%, respectively. The passivated planar films were easily prepared by direct gas/solid reaction of a hydrohalide-deficient lead precursor with CH_3NH_2 gas. The amount of PbI_2 impurity was controlled via the hydrohalide deficiency of the precursor. The PbI_2 was expected to form during crystallization, rather than existing in precursor films or in the as-grown perovskite film.

Room-temperature pure-violet light-emitting diodes, based upon 2-dimensional 2-phenylethylammonium lead bromide, $(C_6H_5CH_2CH_2NH_3)_2PbBr_4$, were described[240]. The natural quantum-confinement afforded by a 2-dimensional layered perovskite permitted a photoluminescence of shorter (410nm) wavelength than did the 3-dimensional counterpart. As-deposited polycrystalline thin film was converted into micron-sized nanoplates by solvent vapour annealing and integrated into light-emitting diodes. This led to efficient room-temperature violet electroluminescence at 410nm, with a narrow bandwidth.

Early measurements of the heat of formation of $CH_3NH_3PbI_3$ showed[241] it to be thermodynamically unstable with respect to decomposition into lead iodide and methylammonium iodide. This occurred even in the absence of ambient air, light or heat-induced defects, and was assumed to limit its long-term use in practical devices. The formation enthalpies of binary halides in fact become less favorable in the order: $CH_3NH_3PbCl_3$, $CH_3NH_3PbBr_3$, $CH_3NH_3PbI_3$. The chloride alone has a negative heat of formation. Geometrical matching of the ion-sizes of constituents can however increase stability.

Fluorescence polarization was studied[242] in $CH_3NH_3PbI_3$, $CH_3NH_3PbI_{3-x}Cl_x$ and $CH_3NH_3PbBr_3$ perovskites which had been prepared using various methods. Testing was performed on samples with and without a hole-transport layer. The $CH_3NH_3PbI_3$ exhibited an isotropic fluorescence polarization, while that of the others was anisotropic. The composition markedly affected this anisotropy, whereas the preparation method had a relatively weaker effect. The presence of a hole-transport layer could accelerate decay of the fluorescence anisotropy. It was concluded that the symmetry of the perovskite structure was the main factor affecting fluorescence anisotropy.

By systematically modifying the electron-selective contacts and the thickness of the perovskite film, it is possible to distinguish between the impedance signals due to the perovskite layer and those arising from the contacts. Such a study[243] was carried out using a mixed organic lead halogen perovskite $[CH(NH_2)_2]_{0.85}(CH_3NH_3)_{0.15}Pb(I_{0.85}Br_{0.15})_3$ solar cell and SnO_2, TiO_2 or Nb_2O_5 electron-selective contacts. The atomic-layer-deposited contacts were pinhole-free, and of controlled thickness. It was found that the interfacial impedance had a rich structure involving various capacitive processes, several electron-extraction steps,, and a prominent inductive loop which was related to a negative capacitance at intermediate frequencies.

Trap states and other factors cause carrier-loss in organometallic perovskite solar cells due to carrier recombination at the surface and in the sub-surface of the perovskite[244]; thus leading to an impaired conversion efficiency. One means of reducing carrier losses is to introduce an interfacial layer. Trap states at the perovskite surface are then efficiently suppressed, leading to an homogenous surface potential which avoids surface carrier-recombination. The Fermi level of a perovskite can be shifted toward the valence band by perhaps 0.2eV; thus introducing an energy barrier to electron diffusion, and minimizing carrier recombination in the perovskite sub-surface. The performance of the cell then markedly improves, with the average efficiency increasing from 14.3 to 16.4%, and with a maximum efficiency of 18.1%. There is also the bonus that non-encapsulated cells having a dual-function interfacial layer exhibit increased long-term stability in ambient air.

The first high-pressure study of single-crystal hybrid perovskites of the form, $CH_3NH_3PbX_3$ (X = Br or I), revealed[245] a striking piezochromism in which the solids first became lighter in colour and then went black under compression. Electronic conductivity measurements of $CH_3NH_3PbI_3$ in a diamond-anvil cell showed that the resistivity decreased by 3 orders-of-magnitude between 0 and 51GPa. The activation energy for conduction at the latter pressure was only 13.2meV and this suggested that the perovskite was approaching a metallic state. It had been reported that $CH_3NH_3Pb(Br_xI_{1-x})_3$ (x = 0.2 to 1) perovskites reversibly formed light-induced trap states which pinned the

Materials Research Forum LLC
https://doi.org/10.21741/9781644900819

photoluminescence to a low energy. The newer high-pressure data indicated that compression could mitigate such a photoluminescence red-shift.

It has been observed[246] that oxygen can, with the assistance of laser excitation, be intercalated into the framework of $CH_3NH_3PbI_3$. Intercalated oxygen can be easily removed by reducing the ambient pressure. Studies suggested that Pb-O bonds formed mainly at the surface of the $CH_3NH_3PbI_3$ but are somehow prevented from forming PbO.

Three-dimensional TiO_2 having a hierarchical architecture which was based upon rutile nanorods was essayed[247] as a photo-anode material for perovskite solar cells. These nanorod films were prepared by using a 2-step low-temperature hydrothermal method at 180C and consisted of TiO_2 nanorod trunks, with optimum lengths of 540nm, and TiO_2 nanobranches with lengths of 45nm. Solar cells which were based upon hierarchical nanorods and $CH_3NH_3PbI_3$ led to the highest power-conversion efficiency. This was attributed mainly to lower charge-recombination. When the $CH_3NH_3PbI_3$ was deposited by using the 2-step method, it led to greater efficiency and surface coverage than did 1-step preparation of $CH_3NH_3PbI_{3-x}Cl_x$.

The incorporation of PbI_2 into perovskite light-absorbers in solar cells had a beneficial effect[248] in that it led to a reduced hysteresis and ionic migration and to a power-conversion efficiency of up to 19.75% under an illumination intensity of $100mW/cm^2$.

The data arising from early electrical measurements performed on $CH_3NH_3PbI_3$ photovoltaic cells could be quantitatively explained[249] in terms of established models for inorganic semiconductors. The current-voltage relationship could be well explained in terms of a 2-diode model. Impedance spectroscopy meanwhile revealed the presence of a thick intrinsic layer. It was concluded that $CH_3NH_3PbI_3$ is an ideal intrinsic semiconductor which is very resistant to the effects of accidental doping.

The main cause of photo-induced degradation in solar-cells having mesoporous and planar structures was proved[250] to be related to the interface between the hole-transport layer and the metal electrode. Solar irradiation caused an appreciable fall in performance, with the efficiency decreasing from about 18 to 2.46% within 3h, in the case of planar cells. The deterioration was attributed to retarded carrier-extraction, from the hole-transport layer to the gold electrode, and this in turn was due to broken interface bonding. *In situ* renewal of the gold electrode led to recovery of some 80% of the performance in the case of both mesoporous and planar cells. The $TiO_2|$perovskite interface had little effect upon photo-induced degradation.

Layers of $CH_3NH_3PbI_3$ for use as light-absorbing material were prepared[251] by annealing spin-coated PbI_2 thin films in CH_3NH_3I vapour. The required perovskite began to form at above 140C, with signs of PbI_2 completely disappearing at above 160C. The grain-size also increased with increasing annealing temperature.

Thin films of nanocrystalline $CH_3NH_3PbI_3$, having various concentrations of methylammonium iodide, were deposited onto glass substrates by means of single-step spin-coating. It was noted[252] that PbI_2 was present, even for stoichiometric atomic concentrations of the main components. The PbI_2 was highly textured along the (001) direction, with the crystallite size ranging from 30 to 40nm. The optical-absorbance edge of the $CH_3NH_3PbI_3$ thin films could be described in terms of the Tauc direct-transition model. The calculated band-gap of the pure PbI_2 film was 2.40eV, while the calculated band-gap of perovskite film having a stoichiometric ionic ratio of PbI_2 and CH_3NH_3I was 1.46eV, with an absorption coefficient of some 10^5/cm. There occurred a red-shift in the perovskite emission with increasing methylammonium iodide concentration.

The growth of monocrystalline lead halide perovskite arrays onto a poly(3,4-ethylene dioxythiophene) polystyrenesulfonic acid coated indium tin oxide substrate, by means of droplet-pinned crystallization, led[253] to a power-conversion efficiency of 1.73%.

Devices having the configuration, ITO/PEDOT:PSS|CH$_3$NH$_3$PbI$_3$|PCBM|metal, when subjected to 4h of steady illumination exhibited[254] a marked decrease in power-conversion efficiency, from 12 to 1.8%. This was traced to chemical degradation of the metal contact. The use of a Cr_2O_3|Cr contact increased stability, but there was a change in the energetic profile during steady illumination which again impaired the performance. The properties of the bulk perovskite layer were little affected by the degradation process. The electrical properties of the cathode contact were modified by the generation of a dipole, at the cathode, which caused a large shift of the flat-band potential that changed the interfacial energy barrier and impeded the efficient extraction of electrons. Ionic motion in the perovskite layer changed the energy-profile close to the contacts and affected energy-level stabilization at the cathode.

Current hysteresis was soon linked[255] to the kinetics of ion migration and local rearrangements of ions at the electrode interfaces gave rise to capacitive effects. Charging transients in thick $CH_3NH_3PbI_3$ samples, following step-changes in voltage, involved 2 main polarization processes and reflected the existence of an ionic double-layer at the interface with the non-reacting contacts. Ionic charging, having a response time of about 10s, was an effect which was very much confined to the vicinity of the electrode. This implied the absence of a net mobile ionic concentration in the bulk material.

[256]Time-resolved photoluminescence spectroscopy reveals that the bulk defects in perovskites which are prepared by means of vapour-assisted solution processing play an important role. The defect density ($\sim 10^{17}$/cm^3) of material which was prepared at 150C

was found to increase 5-fold at 175C, although the average grain-size slightly increased. This proved that grain-boundary defects were not the main mechanism which was responsible for observed differences in photoluminescence behaviour as a result of annealing. Following surface passivation using water molecules, the photoluminescence intensity and lifetimes of samples which were prepared at 200C improved to some extent but were still much lower than those which were prepared at 150C. It was therefore concluded that most defect states observed in material which was grown at high temperatures were due to bulk defects.

Density functional theory in the local-density approximation was used[257] to analyze methylammonium lead iodide with regard to its electronic structure and light-absorption properties. Because the nature of the excitons in these materials is close to being of Wannier-Mott type, the simulation of solar cells involved a drift diffusion model An unexpected narrow intermediate band was created, by mercury-doping, in the forbidden band-gap at about 0.5eV above the valance band. The material was predicted to have a wide absorption band in the solar spectrum.

First-principles calculations, performed within density functional theory and taking account of spin-orbit coupling, were performed[258] for $PbCl_2$, $PbBr_2$ and $CH_3NH_3PbBr_{3-x}Cl_x$. When a suitably modified Becke-Johnson exchange potential was used, there was very good agreement with experiment. The method was subsequently applied to predicting the effect of replacing the methylammonium cation, in $CH_3NH_3PbI_3$ or $CH_3NH_3PbBr_3$, with $N_2H_5^+$ or $N_2H_3^+$ cations, which possess slightly smaller ionic radii. It was predicted that an appreciable decrease in the band-gap would result from replacement and that the effect would be to extend optical absorption down to the near-infrared.

It was found that, by using so-called photonic flash sintering, a spin-coated perovskite precursor could be fully annealed in as little as 1ms[259]. Devices which were annealed within 1.15ms exhibited power-conversion efficiencies of up to 11.3%, as compared to the 15.2% efficiency which was found for samples which had been annealed for 1.5h on a hot-plate.

Perovskite films which consisted of atomically smooth monocrystalline domains of up to 5 microns in size could be prepared[260] by the judicious annealing of the solvent and precursor components. The defect density was halved, as compared with that of thermally-annealed film. Improvements in crystallinity and the defect states led to a 53% increase in solar-cell efficiency; with the latter attaining 14.13%.

A simple deposition-method which was based upon gas-quenching was shown[261] to produce smooth pinhole-free films of materials ranging from $CH_3NH_3PbI_3$ to triple-cation perovskites. It could also produce planar heterojunction devices having high efficiencies.

The morphology and exposed facets of solution-processed $CH_3NH_3PbI_3$ thin films, used as the active-layer material in solar cells, contribute[262] to efficiency and stability. When growing $CH_3NH_3PbX_3$ (X = Br or I) crystals from solution, bromide, non-aqueous solvents and increased X/PbX_2 ratios encourage the growth of cuboids rather than dodecahedra. These factors promoted the coordination of X^- to Pb_2^+ on the growing surfaces and thereby made the {110} faces out-grow the {100} faces.

A sandwich-structure has been proposed which does not require electron-transport or hole-transport layers, or exotic electrodes. Fluorine-tin-oxide|$CH_3NH_3PbI_3$ and fluorine-tin-oxide|$CH_3NH_3PbBr_3$ thin films, prepared[263] by means of solvent engineering and halide exchange, respectively, were stacked perpendicularly facing each other with the glass surface outwards. The assembly was then encapsulated in epoxy. The photocurrents and sensitivities of $CH_3NH_3PbI_3$-$CH_3NH_3PbBr_3$ heterojunctions could be improved by an order-of-magnitude.

The methylammonium materials, $CH_3NH_3PbI_3$ and $CH_3NH_3Pb(I_{0.95}Br_{0.05})_3$, were prepared[264] as thin films deposited onto micro-structured gold electrode arrays on Si/SiO_2 wafers. Hysteresis occurred in the I-V characteristics of all of the samples, whether in the dark or under illumination. Persistent changes in the polarization of the perovskite films appeared, after positive or negative poling, leading to appreciable changes in the current density and residual current, in the absence of an applied bias. At higher bias voltages, new inverted hysteresis loops appeared which indicated a decrease in the barrier width and/or height at the perovskite|metal contact. The I-V characteristics in that voltage range could be modelled as 2 back-to-back diodes.

The spectral dependence of direct and trap-mediated decay processes in $CH_3NH_3PbI_3$ was determined using time-resolved microwave conductivity[265]. The total end-of-pulse mobility depended upon the excitation wavelength. The maximum (172cm^2/Vs) mobility occurred when the charge-carriers were excited by using near-bandgap light (780nm) in the low (10^9photons/cm^2) charge-carrier density regime, and was lower for above-bandgap and sub-bandgap excitations. Direct recombination occurred, at a 100 to 400ns timescale, for excitation wavelengths which were near to and above the bandgap. Indirect recombination processes exhibited differing behaviours following above-bandgap and sub-bandgap excitations, suggesting the influence of different trap distributions on recombination dynamics.

The optical behaviour of $CH_3NH_3PbI_3$ is governed mainly by surface states, and lead dangling bonds on the surface introduce shallow electronic states[266]. The associated carrier-localization effects of the electronic states are relatively weak: the lifetimes of carriers on the iodine-poor surface are comparable to those in the interior. Iodine-rich areas on the surface introduce deep trap centers for carriers, and these reduce carrier diffusion lengths. Surface passivation prolongs carrier diffusion lengths, and acts mainly on enriched-iodine on the surface rather than on lead dangling bonds.

Polaronic exciton binding energies, deduced from permittivity data, were consistent[267] with experimental energies for $CH_3NH_3PbI_3$ and $CH_3NH_3PbBr_3$ over a wide temperature range. The band-gaps exhibited a discontinuity at the orthorhombic-to-tetragonal transition of the iodide, but not of the bromide.

Another growth method combined the 2-step process with chlorine incorporation, and solar cells made from the resultant material exhibited[268] a power-conversion efficiency of 10.5%; 27% better than that without chlorine incorporation. Kelvin-probe force microscopy revealed greater band-bending at grain boundaries when chlorine was incorporated, and this brought the Fermi level of the bulk perovskite thin film closer to the center of the band-gap. As a result, p-i-n type junctions were formed in devices with incorporated chlorine, which aided charge-carrier collection. The electron lifetime in perovskite thin films with chlorine was also longer, thus indicating reduced recombination in devices containing chlorine.

Exciton binding-energy calculations were based[269] upon an f-sum rule for integrated visible ultraviolet absorption spectra. Between 80 and 300K, the exciton binding energy in $CH_3NH_3PbBr_3$ was 60meV regardless of the temperature. For $CH_3NH_3PbI_3$ in the orthorhombic phase (<140K) the binding energy was 34meV. In the tetragonal phase, the energy decreased to 29meV at 170K and then remained constant up to 300K.

Predictions were made of the key characteristics of defect-free $CH_3NH_3PbI_3$-based solar cells having various materials, including spiro-OMETAD, NiO, CuI, CuSCN or Cu_2O as hole-transport layers[270]. The results showed that cells which used Cu_2O as the hole-transport layer were the best, with a power-conversion efficiency of more than 24%.

Short carrier lifetimes arising from electron-hole bi-molecular recombination, and corresponding to the fast decay of photoluminescence, were found[271] in $CH_3NH_3PbI_3$ films which were annealed at high temperatures. The doping nature of the perovskite varied from p-type, via intrinsic to n-type with increasing annealing temperature. The short carrier-lifetime and the intrinsic nature of the perovskite led to a high open-circuit voltage for corresponding cells. On the other hand, n-type doping led to a high photocurrent and efficiency. By adjusting the carrier lifetime and doping behaviour, cells having power-conversion efficiencies of better than 17% could be prepared.

Films of $CH_3NH_3PbI_3$ which were grown by co-evaporating CH_3NH_3I and PbI_2 were compared[272] with the $CH_3NH_3PbI_3$ and $CH_3NH_3PbCl_3$ films which resulted from evaporation using CH_3NH_3I and $PbCl_2$ precursors. The main differences were the room-temperature crystal structures. The preferred orientation of pure $CH_3NH_3PbI_3$ depended upon the precursor molar flux-ratio which was used, and could also be modified by annealing.

Lead halide thin films of various thickness, having a porous morphology and low crystallinity, could be produced[273] by adding $PbCl_2$ powder to a PbI_2 solution of DMF precursor solution. Planar cells, based upon 300nm-thick $CH_3NH_3PbI_{3-x}Cl_x$ thin film made from precursor solutions of a mixture of 0.80M PbI_2 and 0.20M $PbCl_2$, exhibited a power-conversion efficiency of 10.12% together with an open-circuit voltage of 0.93V, a short-circuit photocurrent density of 15.70mA/cm^2 and a fill-factor of 0.69.

A mesoporous $CH_3NH_3Pb(I_{0.88}Br_{0.12})_3$ solar cell with a power-conversion efficiency of about 14% exhibited[274] appreciable potential-barrier bending at grain boundaries, and an induced passivation. The potential difference value for a bromine-free cell was about 50mV, and the distribution of the positive potential was lower than that for the x = 0.12 material. Charged grain boundaries had a beneficial effect upon electron-hole de-pairing and in suppressing recombination, thus improving the efficiency.

It has been pointed out that capacitance is a key parameter for investigating the mechanisms operating in these materials, especially those related to current hysteresis[275]. The dielectric properties of microscopic dipolar units account for the intermediate-frequency capacitance. Electrode-polarization, caused by interfacial effects and assumed to be related to kinetically slow ions which pile up near to the outer interfaces, reliably explains excess capacitance values at low frequencies. The current-voltage curves and capacitive responses of perovskite-based solar cells are connected, and hysteresis in dark currents is attributed to slow capacitive mechanisms.

A photo-induced dielectric constant of about 10^6 has been reported[276] for lead halide perovskite solar cells, and similar effects have been found in measurements of porous lead zirconate titanate samples, saturated with water. The main effect of illuminating a solar cell and of introducing water into the pores of lead zirconate titanate is a marked increase in conductivity and dielectric loss, leading to a low-frequency power-law dispersion. Use of the Kramers-Kronig relationship shows that the large permittivities are related to the power-law changes in conductivity and dielectric loss, and those power-laws are consistent with an electrical-network model for the microstructure. It was concluded that the high permittivity values were artefacts of the microstructural network rather than being fundamental effects.

Single-crystal nanowires, nanorods and nanoplates of $CH_3NH_3PbI_3$ and $CH_3NH_3PbBr_3$ were grown[277] via dissolution-recrystallization from lead iodide or lead acetate films coated onto a substrate. These nanostructures exhibited a strong room-temperature photoluminescence and long carrier lifetime. Solid-liquid interfacial reaction could produce a highly crystalline nanostructured $CH_3NH_3PbI_3$ film with a micron grain-size and high surface coverage. That then led to devices having a power-conversion efficiency of 10.6%.

Evidence was found[278] for energy-level alignment at hybrid interfaces between lead halide perovskite and organic hole-transport materials. The alignments found at perovskite|hole-transport-layer interfaces involved 4 entirely different energy-level offsets depending upon whether the hole-transport material was spiro-OMeTAD, NPB, F16CuPc, HATCN or MoO_3. It was concluded that staggered-gap heterojunction contact with a hole-transport-material higher-lying occupied molecular orbital can aid interfacial hole extraction.

An all solution-processed tandem water-splitting assembly has been described[279] which comprised a $BiVO_4$ photo-anode and a single-junction $CH_3NH_3PbI_3$ hybrid perovskite solar cell. This combination achieved efficient solar-photon collection, with the metal-oxide photo-anode selectively harvesting high-energy visible photons and the perovskite solar cell capturing lower-energy visible to near-infrared wavelengths all at once. When operating with no external bias under standard illumination, this apparatus – plus a suitable catalyst – could attain a solar-to-hydrogen conversion efficiency of 2.5%.

Sequential fabrication was used to create planar $CH_3NH_3PbI_3$ solar cells with a CuSCN hole-transport layer[280]. In the PbI_2 layers, made by spin-coating, small amounts of CH_3NH_3I and dimethyl sulfoxide were first incorporated as the precursor layer on a flat TiO_2 surface. As the first-drip precursor layer, CH_3NH_3I solution was applied by either soaking or dripping using successive spin-coating. When using the normal sequential CH_3NH_3I-soaking method, it was not possible to prepare planar $CH_3NH_3PbI_3$ perovskite solar cells with a CuSCN hole-transport layer. On the other hand, by using the CH_3NH_3I-dripping method a significant photovoltaic effect was observed in planar $TiO_2|CH_3NH_3PbI_3|CuSCN$ solar cells.

Attempts have been made to replace spiro-OMETAD as the hole-transport layer[281]. Required properties such as efficiency, fill-factor, open-circuit voltage and short-circuit current were calculated for cells which had CuI, CuSCN, NiO or Cu_2O as the hole-transport material. A defect-free n-$TiO_2|CH_3NH_3PbI_3|$p-Cu_2O structure offered the highest efficiency, of about 25%, but this was reduced to 22% in the presence of a defect located at 0.45eV above the valance band of the Cu_2O. A high open-circuit voltage of 1.13eV for the p-Cu_2O-based structure indicated a minimal energy-loss due to charge-transfer across the hetero-junctions. Another possible replacement was suggested[282] to be the inorganic p-type hole-transport material, copper thiocyanate. Using a low-temperature solution-deposition method, a power-conversion efficiency of 12.4% was achieved. A further possible replacement was suggested[283] to be copper iodide. By using this, a power-conversion efficiency of 6.0% was obtained together with very good photocurrent stability. The open-circuit voltage however remained low when compared with the best spiro-OMeTAD devices, and this was attributed to a higher recombination in CuI. The latter however had a 2 orders-of-magnitude higher electrical conductivity than that of spiro-OMeTAD, with much higher fill-factors. As well as searching for alternatives to spiro-OMeTAD, there has also long been an enthusiasm for finding alternative electrode materials[284]. The proposed alternatives to thermally evaporated silver and gold electrodes included solid-state dye-sensitized solar cells and organic photovoltaics which were based upon the use of solution-processed silver nanowires. Another innovation was the suggestion that a highly-conducting self-adhesive laminate electrode could be applied to devices.

The ferro-electric properties of $CH_3NH_3PbI_3$ were studied[285] by means of piezo-electric force microscopy and macroscopic polarization measurements. Electric polarization was clearly indicated by amplitude and phase hysteresis loops, but the polarization loop decreased with decreasing frequency and persisted for the order of only 1s. This indicated that $CH_3NH_3PbI_3$ does not exhibit permanent polarization at room temperature. A marked increase in the piezo-electric response occurred under illumination and was consistent with a reported giant photo-induced dielectric constant at low frequencies. The effect was attributed to an intrinsic charge-transfer photo-induced dipole in the perovskite cage.

A giant dielectric-constant phenomenon is observed in these materials, which takes the form of a low-frequency dielectric constant - in the dark - of the order of 1000. This constant can increase further under illumination or due to charge-injection under an applied bias[286]. The constant increases almost linearly, with illumination intensity, by up to a factor of 1000. The giant dielectric constant was considered to be an intrinsic property of $CH_3NH_3PbX_3$ and was attributed to structural fluctuations. Photo-induced carriers modified the local unit-cell equilibrium and changed the polarizability; helped by the freedom-of-rotation of CH_3NH_3.

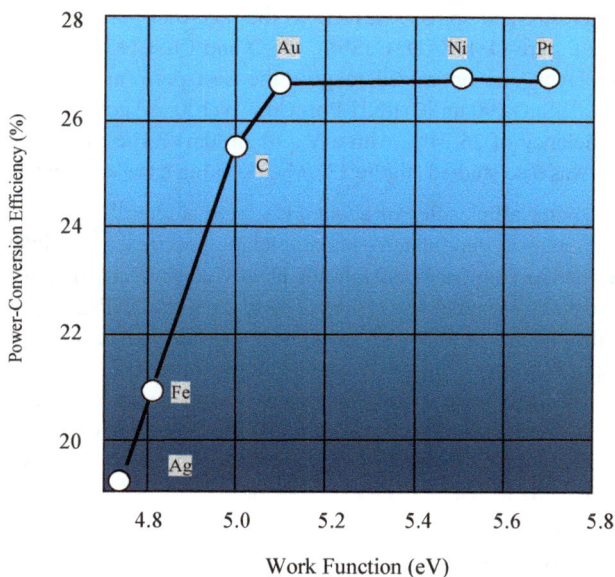

Figure 13. Effect of the work function of the metal contact on the power-conversion efficiency of planar $CH_3NH_3PbI_3$ perovskite solar cells with a SnO_2 electron-transport layer and a CuSCN hole-transport layer

The application of an electric field under inert conditions leads only to a reversible poling which exists for the order of minutes. The presence of moisture, or of any small polar hydrogen-bonded molecule, results in irreversible degradation in the presence of an electric field and occurs within a period of hours under conditions which are relevant to photovoltaic-device operation. It has been suggested[287] that irreversible field-induced degradation in air occurs via an hydrated phase in which a loosely-bound organic cation can drift in an electric field and finally degrade the material to plain PbI_2. Hysteresis in current-voltage curves is aggravated by the presence of moisture. Devices which are aged under load meanwhile exhibit an accelerated degradation. It was shown that ion-migration under inert conditions is largely reversible and occurs via defect-motion. Under humid conditions, significant methylammonium-ion drift leads to irreversible structural degradation.

Replacements for the usual electron-transport materials of solar cells have been considered[288]. The effect of their thickness was also studied for cells having $CH_3NH_3PbI_3$

as the light-harvester and spiro-OMeTAD as the hole-transport material. In an effort to avoid the use of spiro-OMeTAD, $CuSbS_2$, Cu_2O and CuSCN were investigated and the hole-transport layer was optimized to give the best performance. The configuration, fluorine-tin-oxide|SnO_2(90nm)|$CH_3NH_3PbI_3$|CuSCN(100nm)|gold led to a power-conversion efficiency of 26.74%, with a V_{oc} of 1180mV. The role of the metal cathode work-function was also studied (figure 13) when seeking a replacement for gold.

Twelve cells having efficiencies of about 11% were studied[289] by means of impedance spectroscopy, intensity-modulated photovoltage spectroscopy, intensity-modulated photocurrent spectroscopy and open-circuit photovoltage decay methods. Although the cells were essentially identical, the results divided them into 2 distinct groups. Half of the cells had ideality factors of about 2.5 while the other half had ideality factors of about 5. Impedance spectroscopy, performed under illumination and open-circuit conditions, showed that cell capacitance was dominated by the geometrical capacitance of the perovskite layer rather than by a chemical or diffusion capacitance due to photogenerated carriers. The cells with a 2.5 ideality factor exhibited a persistent photovoltage effect that was absent from cells with a higher ideality factor.

Films prepared by the deposition of solutions containing lead halides and the $CH_3NH_3^+$ organic cation produce the $CH_3NH_3PbI_3$ perovskite structure during annealing. It was noted[290] that Pb_2^+ easily forms plumbate complexes in the presence of excess iodide ions, and exhibits characteristic absorption bands at 370nm (PbI_3^-) and 425nm (PbI_4^{2-}). By comparative spectral analysis of the absorption of charge-transfer complexes in the solution phase, and those of the final solid-state perovskite films, it was possible to classify completely the absorption features of the excited state of $CH_3NH_3PbI_3$ across the transient absorption spectrum which resulted from laser-pulse excitation. A broad photo-induced absorption was attributed to a charge-transfer excited state, and there was found to be a correlation between the photo-induced absorption and 480nm bleach signals.

Optical spectroscopy and multi-scale modelling showed that electron-hole interaction is sensitive to the microstructure of the material. Long-range order is disrupted by polycrystalline disorder, and variations in the electrostatic potential occurring in smaller crystals suppress exciton formation, while larger crystals having the same composition exhibit a clearly excitonic state[291].

It was suggested[292] that the rapid reaction kinetics of PbI_2 with CH_3NH_3I permitted the pre-formation of the perovskite within the intermediate film matrix, ready for the subsequent growth of $PbCl_2$-derived perovskite. As-prepared perovskite film had a textured crystal morphology and better compactness. This promised excellent light-absorption and long-term retention of photogenerated charge-carriers. This led to an 11%

efficiency of related planar-solar cells. The film morphology was also very stable during long-term heating.

The addition of chloride, and sequential deposition, were combined[293] into a sequential method by spin-coating $PbCl_2$ plus PbI_2 onto a mesoporous TiO_2 film; followed by transformation to the perovskite. The role played by chlorine in determining the structural and morphological properties was related to the photovoltaic performance. The highest efficiency, for a 10mol%$PbCl_2$ addition, was 14.15%; with an open-circuit voltage of 1.09V, a fill-factor of 0.65 and a short-circuit current of 19.91mA/cm^2. The favorable results were attributed to an increase in the film conductivity, due to a better perovskite morphology.

In the case of $CH_3NH_3PbI_3$ and $CH_3NH_3PbBr_3$ it was again found[294] that the addition of $PbCl_2$ to the solutions used during preparation had a striking effect upon the product, because $PbCl_2$ nanocrystals, present during preparation, acted as heterogeneous nucleation sites for the formation of perovskite crystals in solution. Optimization of the creation of solar cells based upon the lead bromide perovskite led to an efficiency of 5.4% and an open-circuit voltage of 1.24V.

Density functional theory calculations were used[295] to predict how modes of atomic layering could affect the near-gap electronic structure of tin- and lead-halide perovskites, with a view to dye-sensitized solar cells. It was found that, regardless of how the atomic layering was done, the band-gaps could always be widened by several tenths of an electron-volt or more. Most of the trends in the band-gaps were related to the effects of atomic layering and quantum-confinement upon the nature and energy of crystal orbitals.

The macroscopic electric field distribution, in planar structures comprising $CH_3NH_3PbI_{3-x}Cl_x$, was determined[296] by using capacitance-voltage and Kelvin-probe techniques. The device had a p-doped character and formed a p-n heterojunction with n-doped TiO_2 compact layers. The depletion-width at equilibrium was about 300nm; about half of the layer-thickness. There was therefore an appreciable neutral zone adjacent to the anode contact.

The organic-inorganic hybrid perovskites, $CH_3NH_3PbI_{3-x}Cl_x$ and $CH_3NH_3PbI_3$, were deposited from solution onto mesoporous TiO_2/glass substrates. A long-term transition from disordered to more crystalline was observed[297] in both materials over a period of some 8400h. The rate of this conversion could be hastened by increasing the sample temperature. The 2 materials had very different surface microstructures; both were dominated by a similar crystal structure but the preferred orientations were different. The results suggested that the material described as being $CH_3NH_3PbI_{3-x}Cl_x$ was actually a combination of $CH_3NH_3PbI_3$ and $CH_3NH_3PbCl_3$.

When $CH_3NH_3PbInBr_{3-x}$ (x = 0 to 3) was considered[298] as a hole conductor and light-harvester in solar cells, various concentrations of methylammonium iodide and methylammonium bromide were studied, revealing that any composition would conduct holes. The hybrid perovskite was deposited in 2 steps, using 2 precursors separately in order to permit better control of the composition and the related band-gap. Hybrid iodide/bromide hole-conductor free solar cells exhibited very good stability, with a power-conversion efficiency of 8.54% and a current density of $16.2mA/cm^2$.

Organic lead halide perovskite $CH_3NH_3PbI_{3-x}Cl_x$ solution could be used[299] as a liquid electrolyte in dye-sensitized solar cells. The presence of inorganic octahedra, PbX_6^{4-} (X = I or Cl), in the perovskite solution could markedly improve device stability and enhance the photo-response.

The effect of electron-selective and hole-selective contacts on the behaviour of $CH_3NH_3PbI_3$ solar cells was studied[300] by means of impedance spectroscopy. Cells having compact TiO_2 and spiro-OMeTAD as electron-selective and hole-selective contacts were compared with cells which lacked one of those contacts. The contacts increased the fill-factor, and the hole-selective contact was largely responsible for a high open-circuit voltage. The recombination-rate was also determined mainly by the selective contacts.

Treatment, with thiophene or pyridine, of perovskite crystal surfaces was shown[301] to reduce markedly the incidence of non-radiative electron-hole recombination in $CH_3NH_3PbI_{3-x}Cl_x$, thus leading to photoluminescence lifetimes which were increased to 2μs; almost an order-of-magnitude increment. This was attributed to the electronic passivation of under-coordinated lead atoms in the crystal. Due to the Lewis-base passivating effect of the thiophene or pyridine, power-conversion efficiencies in solution-processed planar heterojunction solar cells could be increased from 13%, for untreated cells, to 15.3 for thiophene-treated and to 16.5% for pyridine-treated cells.

An important factor when preparing perovskite-based solar cells is that the deposition conditions must ensure complete coverage, together with an adequate film thickness, as these affect other properties. In 2-step deposition, the current density is affected by changing the spin velocity, and the fill-factor changes mainly with dipping-time and methylammonium halide concentration. The open-circuit voltage however is essentially unaffected by those parameters[302].

Uniform fully-converted $CH_3NH_3PbI_3$ films were created[303] at room temperature by carrying out the conversion of PbI_2 in a saturated solution of CH_3NH_3I in cyclohexane. This method produced films having a 20nm root-mean-square roughness.

A vacuum-based co-evaporation method, involving 2 sources, was used[304] to grow $CH_3NH_3Pb(I,Cl)_3$ perovskite thin films. The use of different $CH_3NH_3I/PbCl_2$ flux ratios led to the creation of 2 distinct $CH_3NH_3Pb(I_xCl_{1-x})_3$ phases; one with a high (x > 0.95) iodine content and one with a low (x < 0.5) iodine content. A wide intermediate miscibility-gap existed. The perovskite film decomposed into PbI_2 at temperatures above 200C.

A temperature-dependent study of vapor-deposited $CH_3NH_3PbI_{3-x}Cl_x$ showed that, during cooling, there was an up-shift at 100K in the onset of absorption, by about 0.1eV. This was attributed[305] to tetragonal-orthorhombic transformation of the pure $CH_3NH_3PbI_3$. During further temperature decreases, a second photoluminescence emission peak appeared, in addition to the peak arising from the room-temperature phase. The transition, during heating, occurred at about 140K, thus revealing the existence of significant hysteresis. The photoluminescence-decay lifetimes were independent of temperature when above the transition, but greatly accelerated recombination occurred in the low-temperature phase. It was suggested that small inclusions of material which retained the room-temperature phase were responsible for the behavior, rather than a spontaneous increase in intrinsic rate constants, and it was noted that even rare lower-energy sites could affect the overall performance.

A depleted hole-conductor free $CH_3NH_3PbI_3|TiO_2$ heterojunction solar cell was described[306] which incorporated a thick $CH_3NH_3PbI_3$ film. The latter formed large crystals which functioned as both light-harvesters and hole-transport materials. Capacitance versus voltage measurements revealed the existence of a depletion region which extended to both the n and p sides, and the built-in field of that region assisted charge-separation and suppressed back-reaction of electrons from the TiO_2 film to the $CH_3NH_3PbI_3$ film. This solar cell had a power-conversion efficiency of 8%, with a current-density of $18.8mA/cm^2$.

The Hysteresis Problem

When perovskite solar cells were studied[307] using transient photovoltage decay, transient photoluminescence and impedance spectroscopic methods, a slow dynamic process was detected which was associated with characteristic times that were of the order of seconds to milliseconds. These times were related to a novel slow dynamic process which involved some unusual structural properties of the lead halide perovskites such as crystal size and organic cations. This new process was suggested to be a cause of the current-voltage hysteresis, while a low-frequency characteristic time which was commonly associated with the electronic-carrier lifetime could not be attributed to a recombination process. It was not easy however to attribute the hysteresis to just one of several different components, such as the bulk of the perovskite or the various heterojunction interfaces. Among the organo-lead halide perovskites, $CH_3NH_3PbI_3$ was known to have ferroelectric properties, and there was a strong correlation between the transient ferroelectric polarization of $CH_3NH_3PbI_3$, induced by an external bias in the dark, and an increase in hysteresis[308]. Reverse bias poling (-0.3 to -1.1V) of $CH_3NH_3PbI_3$ layers before photocurrent-voltage measurements produced a marked hysteresis whose extent was greatly changed by the cell architecture. This was attributed to the effect of remanent polarization, in the perovskite film, upon the photocurrent; which was most increased in planar perovskites without a mesoporous scaffold. When the hysteresis behavior was studied[309] in planar organic-inorganic lead halide perovskite solar cells which had PC60BM as the cathode, the room-temperature devices exhibited apparently hysteresis-free scans. Cooling to 175K led to the appearance of a marked hysteresis. Chrono-amperometric measurements revealed that the half-time of the relaxation process underlying this hysteresis changed from 0.6s at 298K to 15.5s at 175K; indicating an activation energy of 0.12eV. Cooling of the cell to 77K, under positive bias, permitted – in effect - freezing the cell into the best condition to exhibit efficient photovoltaic behaviour. It was concluded that changes which seemed to remove room-temperature hysteresis might not remove the underlying processes but instead shift them into time-scales which were easily detected by routine room-temperature scans.

The Stability Problem

The most common preparation methods for these materials use solid binary halides as one of the precursors. The binary lead halides exhibit photodecomposition and it was soon recognised that, in view of the perovskite crystal-structure of methylammonium lead halides, it was probable that the lead-halide component would still undergo photodecomposition and a detailed mechanism was proposed[310]. This indicated that the trapping of photogenerated electrons on lead ions in these perovskites should be studied. A study of the effect of ambient humidity upon the crystallization and surface morphology of one-step spin-coated perovskite films, involving experimental analyses and thin-film growth-theory, led[311] to the conclusion that the influence of the humidity upon nucleation during spin-coating was very different from its effect upon crystal growth during annealing. During spin-coating, a high nucleation-density induced by a high supersaturation tended to appear under anhydrous circumstances. This resulted in high-coverage layer growth of perovskite films. During the annealing stage, moderate supersaturation favoured the formation of perovskite films having good crystallinity. Films which were spin-coated under low relative-humidity conditions, followed by annealing under high relative-humidity exhibited an increase in crystallinity and an improved device performance. Samples of $CH_3NH_3PbI_3$, $CH_3NH_3PbI_{3-x}Cl_x$ and $CH_3NH_3PbCl_3$ were compared[312] showing that, in spite of their very different morphologies and formation kinetics, the first 2 materials had very similar electronic structures and chemical compositions; no chlorine was found in the final $CH_3NH_3PbI_{3-x}Cl_x$ samples. On the other hand, the chlorine played a very important role during preparation because it affected the formation of crystalline $CH_3NH_3PbI_3$. When the $CH_3NH_3PbI_3$ was exposed to various water, temperature and storage conditions in air or argon, the main result was to decompose the perovskite into PbI_2. This degradation occurred at 100C and was not related only to high humidity; slow degradation to PbI_2 was observed even in argon and other inert atmospheres.

In line with the perceived danger of humidity, a detailed study was made[313] of the effect of exposure to moisture on methylammonium lead halide perovskite film formation. Films which formed in higher-humidity atmospheres possessed a less continuous morphology but exhibited a markedly improved photoluminescence. Film-formation was also faster. Exposure to moisture in the precursor solution, or in the atmosphere during formation, resulted in a greatly improved open-circuit voltage and thus a better overall device performance. Post-treatment of dry films with moisture improved the photovoltaic performance and the photoluminescence. An enhanced photoluminescence and open-circuit voltage implied that the quality of the material was better in films which had been exposed to moisture, and this was traced to a reduction in the trap density in the films. This in turn was attributed to partial solvation of the methylammonium component and to so-called self-healing of the perovskite lattice. *Ab initio* molecular dynamics simulations, and first-principles density functional theory were used[314] to investigate the effects of sunlight and moisture on $CH_3NH_3PbI_3$ perovskites. The *ab initio* molecular dynamics simulations explored the effect of the impact of a few water molecules on the structures of $CH_3NH_3PbI_3$ surfaces having various terminations. This showed that a PbI_2-terminated surface was the most stable, while CH_3NH_3I-terminated and PbI_2-defective surfaces underwent a structural reconstruction which led to the formation of hydrated compounds in a humid atmosphere. A moisture-induced decrease in photo-absorption was closely related to the formation of hydrated species, and hydrated $CH_3NH_3PbI_3{\bullet}H_2O$ and $(CH_3NH_3)_4PbI_6{\bullet}2H_2O$ crystals hardly absorbed visible light. The electronic excitation of bare and water-absorbed $CH_3NH_3PbI_3$ nanoparticles tended to weaken the Pb-I bonds; especially those around water molecules. The maximum decrease in photo-excitation induced bond-order could attain up to 20% in an excited state in which water molecules were involved in the electronic excitation. This confirmed the occurrence of accelerated decomposition of perovskites in the presence of sunlight and moisture. Surface-trap states and electronic disorder in solution-processed $CH_3NH_3PbI_3$ film greatly affect solar-cell performance, and the moisture-sensitivity of photo-active perovskite material limits practical application. Surface modification of films using the solution-based hydrophobic polymer, poly(4-vinylpyridine), passivated under-coordinated surface lead atoms due to the polymer's pyridine Lewis-base side-chains[315]. This removes surface-trap states and non-radiative recombination, and also acts as an electron-barrier between the perovskite and the hole-transport layer. This reduced the incidence of interfacial charge-recombination and led to an improvement, in open-circuit voltage, by 120 to 160mV. A traditional cell which was fabricated under the same conditions could have an open-circuit voltage as low as 0.9V, due to a predominant interfacial recombination processes. The power-conversion efficiency (15%) was higher by 3 to 5% for polymer-modified devices, with an open-circuit voltage of more than 1.05V and hysteresis-free I–V curves. The hydrophobicity of the polymer chain also protected the surface from moisture and improved the stability of non-encapsulated cells. Performance was retained for up to 720h of exposure in an atmosphere with a relative humidity of 50%. Lead chloride and lead acetate trihydrate, $Pb(Ac)_2 \bullet 3H_2O$, precursors were used[316] to prepare perovskite films. The $PbCl_2$ increased the crystallinity but long annealing-times were required. The $Pb(Ac)_2 \bullet 3H_2O$ was used because it offered rapid crystallization, and the hydrated form promised to improve film stability. An optimum, 1:1 ratio, of these precursors led to compact full-coverage perovskite films with a grain size greater than $1\mu m$. The speed of film-formation was also much faster. These films exhibited stability under moisture for several days in an environment with a relative humidity of less than 70%, and resulted in a power-conversion efficiency of 14.77%.

The effect of water intercalation and hydration upon decomposition and ion migration in $CH_3NH_3PbX_3$ was investigated[317] by means of first-principles calculations. Water interacted with PbX_6 and CH_3NH_3 via hydrogen-bonding, with the former interaction gradually increasing while the latter hardly changed in going from X = I to Br to Cl. The thermodynamics indicated that the water exothermically intercalated into the perovskite. The water-intercalated and monohydrate compounds were seemingly stable against decomposition. Water-intercalation reduced the activation energy for vacancy-mediated ion migration. This value normally increased in going from X = I to Br to Cl. It is clear that the hydration of halide perovskites has to be avoided in order to prevent the degradation of solar cells when exposed to moisture. When the first measurements were made[318] of the electrical bias-induced degradation of inverted perovskite solar cells in the dark, in various environments, it was concluded that humidity together with electrical bias resulted in the rapid degradation of $CH_3NH_3PbI_3$ into PbI_2. This degradation started from the edge of the cell, due to the entry of moisture. The monitoring of local ion migration in the degraded and non-degraded regions showed that the formation of PbI_2

could passivate the perovskite by reducing ion migration. The degradation was very non-uniform across different grains. The bias degradation was closely related to the density of mobile ions.

The Toxicity Problem

It was soon pointed out[319,320] that the hybrid $APbX_3$ perovskites, where A was cesium or CH_3NH_3 and X was bromine or iodine, had excellent photovoltaic properties but were also highly toxic. Lead-free alternatives, based upon tin or germanium are unfortunately very unstable. Stable non-toxic double perovskites which are based upon alternating corner-shared AX_6 and BX_6 octahedra, where A is here silver or copper and B is bismuth or antimony, have indirect band-gaps or band-gaps of more than 2eV. Another group of photovoltaic halides is based upon edge-shared AX_6 and BX_6 octahedra and has the formula: $A_aB_bX_{a+3b}$. This group includes Ag_3BiI_6, Ag_2BiI_5, $AgBiI_4$ and $AgBi_2I_7$. These materials could be very stable, and have direct band-gaps ranging from 1.79 to 1.83eV. A solar cell which was based upon Ag_3BiI_6 exhibited a power-conversion efficiency of 4.3%. In connection with the problematic toxicity of the lead content of these materials, simulations have been performed[321] in which variously encapsulated perovskite modules were mechanically damaged by ice-impact (hail) over a range of weather conditions. An encapsulation method which was based upon epoxy resin reduced the lead leakage-rate by a factor of 375, as compared with an encapsulation method which was based upon a glass cover with an ultraviolet-cured resin at the edges. The greater leakage-reduction by epoxy-resin encapsulation was associated with its good self-healing characteristics and higher mechanical strength. The results implied that the risk of lead-leakage in practice could be reduced to an acceptable degree. This question has been analyzed in some detail[322]. With a view to reducing the toxicity of lead in perovskite solar cells, the structural stabilities, and electronic and optical properties of the mixed perovskites, $(CH_3NH_3Pb)_{0.75}B_{0.25}I_3$ where B was magnesium, calcium, strontium or barium, were predicted[323] on the basis of density functional theory. This showed that the introduction of alkaline-earth metals lowered the structural stabilities of these mixed perovskites as compared to that of $CH_3NH_3PbI_3$. The band-gaps were also widened. It was concluded that the band-gaps of the mixed perovskites were determined mainly by the Pb-I-B bond-angles in the ab-plane. The optical absorption coefficients of all of the metal-doped perovskites were also markedly decreased by adding alkaline-earth metals. The non-toxic alkaline-earth metal cations are nevertheless good candidates for lead-replacement because they maintain the charge balance, and also sometimes satisfy the Goldschmidt-rule tolerance factor. Among Mg^{2+}, Ca^{2+}, Sr^{2+} and Ba^{2+}, the latter is the most suitable and leads to the best power-conversion efficiency. The relationship between charge-carrier dynamics and the Ba^{2+} concentration was determined[324] by means of time-resolved photoluminescence. The Ba^{2+}-doped films which could be processed in a moisture-containing environment with a relative humidity of 1.0% were stable. The optimum 3.0mol%Ba^{2+} content increased the power-conversion efficiency of the resultant solar cell from 11.8 to 14.0%. Returning to the above suggestion of antimony as a lead-replacement, antimony-based hybrid materials of the form, $A_3Sb_2I_9$, were studied[325] with A being CH_3NH_3 or cesium. These materials exhibited good absorption properties, and the band-gaps of $(CH_3NH_3)_3Sb_2I_9$ and $Cs_3Sb_2I_9$ were 1.95 and 2.0eV, respectively. Planar hybrid Sb-based solar cells exhibited little hysteresis, and a power-conversion efficiency of 2.04% was attained by a $(CH_3NH_3)_3Sb_2I_9$ perovskite-based solar cell. It is somewhat ironic that a replacement which imparts further useful properties should be the even more toxic element, thallium. Faceted spheroidal nanocrystals of Tl_3PbX_5, where X is chlorine, bromine or iodine, and $TlPbI_3$ nanowires can be prepared[326] by using colloidal methods. It is possible to vary the band-gap by means of halide substitution, and create materials which can absorb strongly between 250 and 450nm. A confinement effect which is observed in the case of Tl_3PbBr_5 nanocrystals suggests that size-tuning is also possible.

The twin issues of non-optimum light collection due to the thinness (350nm) of polycrystalline absorber layers, imposed by the toxicity of the perovskite, have been addressed[327] by positing the embedding of spherical plasmonic nanoparticles of various sizes and compositions into solar cells. Theory suggests that plasmon-enhanced near-field effects and scattering can lead to a photocurrent improvement of up to 7.3% if silver spheres are embedded in the perovskite layer. A greater improvement, of up to 12%, can result by putting silver spheres in the perovskite and aluminium spheres in the hole-transport layer. Such a use of nanoparticles then permits the use of far thinner (down to 150nm), perovskite layers; thus greatly reducing the toxicity.

Lead is not the only dangerous aspect of perovskite use. There is also the use of toxic solvents such as N,N-dimethylformamide in the preparation of precursor inks. Alternative inks for perovskite precursors have been based[328] upon protic ionic liquids involving methylammonium cations and carboxylate anions, mixed with water, alcohol and acetonitrile, and used for the solution-processing of $CH_3NH_3PbX_3$ (X = I, Br). Blends involving methylammonium propionate were deemed to be the most promising candidates with regard to chemical stability, and compatibility with single- and two-step solution-processing. Multi-cation mixed-halide perovskite solar cells, prepared using protic ionic liquids and acetonitrile, led to materials with power-conversion efficiencies of more than 15%. The thermal degradation of formamidinium-based perovskites and halide precursors has been studied[329] in a helium atmosphere, or under vacuum, using a constant heating-rate of 20C/min, or using pulsed-heating under illumination or dark conditions. The amounts of symtriazine, formamidine and hydrogen cyanide which were released greatly depended upon the temperature. Symtriazine was a thermal degradation product at temperatures above 95C. At lower temperatures, only formamidine and HCN were generated. Formamidinium was more resistant to thermal degradation. It is again clear that lead is not the only toxic material which may complicate the practical applications of perovskites.

Bis(sulfanylidene)tungsten, also known as tungsten disulfide, is a non-toxic replacement material having a direct band-gap of 1.8eV[330]. Perovskite solar cells having bis(sulfanylidene)tungsten as the electron transport-layer and spiro-OMeTAD as the hole-transport layer could feasibly exhibit a power-conversion efficiency of 25.96%, a short-circuit current density of 22.06mA/cm^2, an open-circuit voltage of 1.280V and a fill-factor of 91.76%.

It has recently been shown[331] that lead from perovskites can enter the food chain 10 times more effectively than do lead contaminants already present in the environment, and that the lead from perovskites therefore has to be treated particularly carefully. The safety level for the lead content of perovskite-based should be set lower than those for other lead-containing electronic products.

Other Applications

Attention has been restricted here to the consideration of lead halide perovskites as used in solar cells, but they are also likely to find applications in other more active optical devices. Many types of perovskite laser have been proposed, such as photonic crystal lasers and distributed feedback lasers, but they tend to require the use of complex nanoscale lithography. Perovskite thin films exhibit excellent light-absorption and light emission properties, together with a high defect tolerance and high charge-carrier mobility, making them candidates for solution-processed light-emitting applications and on-chip laser sources. Most interest has centred on lasing behaviour under short-pulse excitation because of the intervention of the so-called laser-death phenomenon which is observed after just a few hundreds of nanoseconds of excitation. This means that continuous-wave operation remains problematic. It was observed only in a single-cation perovskite at a specific temperature of 100K, wherein the optical gain was due to small inclusions in a pump-induced crystal phase. This is clearly not a suitable feature in practical applications. Longer continuous-wave operation is possible over a range of 80 to 130K if the methylammonium cation is partially replaced by formamidinium and cesium[332].

A solar-pumped fiber laser can be produced[333] by using all-inorganic cesium lead halide perovskite quantum-dots as a sensitizer. Such quantum-dots offer marked advantages for such applications because of their broad absorption spectra, narrow emission spectra and high quantum-yields. The peak emission-wavelength of the dots can be varied by adjusting the I/Br ratio so as to produce spectral overlap with Nd^{3+} ions in the laser medium. The laser output-power is very sensitive to the peak emission-wavelength of the quantum-dots. The quantum-dots which were used had a quantum-yield of about 65%. The room-temperature laser output from monocrystalline lead halide perovskite nanowires featured[334] a threshold of $220nJ/cm^2$, corresponding to a charge-carrier density of only $1.5 \times 10^{16}/cm^3$, and a Q-factor of about 3600. A kinetic analysis based upon time-resolved fluorescence revealed little charge-carrier trapping in these nanowires, with an estimated lasing quantum-yield of nearly 100%.

References

[1] Bhatt, M.D., Lee, J.S., New Journal of Chemistry, 41[19] 2017, 10508-10527. https://doi.org/10.1039/C7NJ02691A

[2] Ouyang, R., Chemistry of Materials, 32[1] 2020, 595-604. https://doi.org/10.1021/acs.chemmater.9b04472

[3] Xiao, Z., Song, Z., Yan, Y., Advanced Materials, 31[47] 2019, 1803792. https://doi.org/10.1002/adma.201803792

[4] Yin, W.J., Shi, T., Yan, Y., Journal of Physical Chemistry C, 119[10] 2015, 5253-5264. https://doi.org/10.1021/jp512077m

[5] Meyer, E., Mutukwa, D., Zingwe, N., Taziwa, R., Metals, 8[9] 2018, 667. https://doi.org/10.3390/met8090667

[6] Ono, L.K., Qi, Y., Liu, S.F., Joule, 2[10] 2018, 1961-1990. https://doi.org/10.1016/j.joule.2018.07.007

[7] Odabaşı, Ç., Yıldırım, R., Solar Energy Materials and Solar Cells, 205, 2020, 110284. https://doi.org/10.1016/j.solmat.2019.110284

[8] Zhao, D., Chia, E.E.M., Advanced Optical Materials, 1900783, 2019. https://doi.org/10.1002/adom.201900783

[9] Li, W., Long, R., Tang, J., Prezhdo, O.V., Journal of Physical Chemistry Letters, 10[13] 2019, 3788-3804. https://doi.org/10.1021/acs.jpclett.9b00641

[10] Meggiolaro, D., De Angelis, F., ACS Energy Letters, 3[9] 2018, 2206-2222. https://doi.org/10.1021/acsenergylett.8b01212

[11] Ava, T.T., Al Mamun, A., Marsillac, S., Namkoong, G., Applied Sciences, 9[1]

2019, 188. https://doi.org/10.3390/app9010188

[12] Yang, J., Chen, S., Xu, J., Zhang, Q., Liu, H., Liu, Z., Yuan, M., Applied Sciences, 9[20] 2019, 4393. https://doi.org/10.3390/app9204393

[13] Ye, Q., Zhao, Y., Mu, S., Gao, P., Zhang, X., You, J., Science China Chemistry, 62[7] 2019, 810-821. https://doi.org/10.1007/s11426-019-9504-x

[14] Wang, N., Shen, H., Jin, M., Xu, J., Li, H., Tian, T., Zhang, Y., Journal of the Chinese Ceramic Society, 47[7] 2019, 972-982.

[15] Duan, J., Zhao, Y., Wang, Y., Yang, X., Tang, Q., Angewandte Chemie, 58[45] 2019, 16147-16151. https://doi.org/10.1002/anie.201910843

[16] Thumu, U., Piotrowski, M., Owens-Baird, B., Kolenko, Y.V., Journal of Solid State Chemistry, 271, 2019, 361-377. https://doi.org/10.1016/j.jssc.2019.01.005

[17] Fadla, M.A., Bentria, B., Dahame, T., Benghia, A., Physica B, 585, 2020, 412118. https://doi.org/10.1016/j.physb.2020.412118

[18] Kovalenko, M.V., Protesescu, L., Bodnarchuk, M.I., Science, 358[6364] 2017, 745-750. https://doi.org/10.1126/science.aam7093

[19] Yang, D., Cao, M., Zhong, Q., Li, P., Zhang, X., Zhang, Q., Journal of Materials Chemistry C, 7[4] 2019, 757-789. https://doi.org/10.1039/C8TC04381G

[20] Ananthakumar, S., Kumar, J.R., Babu, S.M., Journal of Photonics for Energy, 6[4] 2016, 042001. https://doi.org/10.1117/1.JPE.6.042001

[21] Kandada, A.R.S., Petrozza, A., Accounts of Chemical Research, 49[3] 2016, 536-544. https://doi.org/10.1021/acs.accounts.5b00464

[22] Xu, F., Zhang, T., Li, G., Zhao, Y., Journal of Materials Chemistry A, 5[23] 2017 11450-11461. https://doi.org/10.1039/C7TA00042A

[23] Zhang, T., Zhao, Y., Acta Chimica Sinica, 73[3] 2015, 202-210. https://doi.org/10.6023/A14090656

[24] Zhang, H., Toudert, J., Science and Technology of Advanced Materials, 19[1] 2018, 411-424. https://doi.org/10.1080/14686996.2018.1458578

[25] Li, G., Price, M., Deschler, F., APL Materials, 4[9] 2016, 091507. https://doi.org/10.1063/1.4962351

[26] Zhao, Y., Zhu, K., Chemical Society Reviews, 45[3] 2016, 655-689. https://doi.org/10.1039/C4CS00458B

[27] Barnett, J.L., Cherrette, V.L., Hutcherson, C.J., So, M.C., Advances in Materials

Science and Engineering, 2016, 4126163. https://doi.org/10.1155/2016/4126163

[28] Miyasaka, T., Chemistry Letters, 44[6] 2015, 720-729.
https://doi.org/10.1246/cl.150175

[29] Cui, J., Yuan, H., Li, J., Xu, X., Shen, Y., Lin, H., Wang, M., Science and
Technology of Advanced Materials, 16[3] 2015, 036004.
https://doi.org/10.1088/1468-6996/16/3/036004

[30] Bruening, K., Tassone, C.J., Journal of Materials Chemistry A, 6[39] 2018, 18865-
18870. https://doi.org/10.1039/C8TA06025H

[31] Chen, C.Y., Lin, H.Y., Chiang, K.M., Tsai, W.L., Huang, Y.C., Tsao, C.S., Lin,
H.W., Advanced Materials, 29[12] 2017, 1605290.
https://doi.org/10.1002/adma.201605290

[32] Sutter-Fella, C.M., Li, Y., Cefarin, N., Buckley, A., Ngo, Q.P., Javey, A., Sharp,
I.D., Toma, F.M., Journal of Visualized Experiments, 2017[127] 2017, e55404.

[33] Zhang, H., Li, D., Cheng, J., Lin, F., Mao, J., Jen, A.K.Y., Grätzel, M., Choy,
W.C.H., Journal of Materials Chemistry A, 5[7] 2017, 3599-3608.
https://doi.org/10.1039/C6TA09845B

[34] Ahmadian-Yazdi, M.R., Eslamian, M., Nanoscale Research Letters, 13, 2018, 6.
https://doi.org/10.1186/s11671-017-2430-0

[35] Prathapani, S., More, V., Bohm, S., Bhargava, P., Yella, A., Mallick, S., Applied
Materials Today, 7, 2017, 112-119. https://doi.org/10.1016/j.apmt.2017.01.009

[36] Manseki, K., Ikeya, T., Tamura, A., Ban, T., Sugiura, T., Yoshida, T., RSC
Advances, 4[19] 2014, 9652-9655. https://doi.org/10.1039/C3RA47870J

[37] Nam, J.K., Chun, D.H., Rhee, R.J.K., Lee, J.H., Park, J.H., Advanced Science, 5[8]
2018, 1800509. https://doi.org/10.1002/advs.201800509

[38] Akbulatov, A.F., Luchkin, S.Y., Frolova, L.A., Dremova, N.N., Gerasimov, K.L.,
Zhidkov, I.S., Anokhin, D.V., Kurmaev, E.Z., Stevenson, K.J., Troshin, P.A.,
Journal of Physical Chemistry Letters, 8[6] 2017, 1211-1218.
https://doi.org/10.1021/acs.jpclett.6b03026

[39] Marronnier, A., Roma, G., Carignano, M.A., Bonnassieux, Y., Katan, C., Even, J.,
Mosconi, E., De Angelis, F., Journal of Physical Chemistry C, 123[1] 2019, 291-
298. https://doi.org/10.1021/acs.jpcc.8b11288

[40] Sarritzu, V., Sestu, N., Marongiu, D., Chang, X., Wang, Q., Masi, S., Colella, S.,
Rizzo, A., Gocalinska, A., Pelucchi, E., Mercuri, M.L., Quochi, F., Saba, M.,

Mura, A., Bongiovanni, G., Advanced Optical Materials, 6[10] 2018, 1701254. https://doi.org/10.1002/adom.201701254

[41] Ravi, V.K., Scheidt, R.A., Dubose, J., Kamat, P.V., Journal of the American Chemical Society, 140[28] 2018, 8887-8894. https://doi.org/10.1021/jacs.8b04803

[42] Ma, Z., Liu, Z., Lu, S., Wang, L., Feng, X., Yang, D., Wang, K., Xiao, G., Zhang, L., Redfern, S.A.T., Zou, B., Nature Communications, 9[1] 2018, 4506. https://doi.org/10.1038/s41467-018-06840-8

[43] Ghosh, T., Aharon, S., Shpatz, A., Etgar, L., Ruhman, S., ACS Nano, 12[6] 2018, 5719-5725. https://doi.org/10.1021/acsnano.8b01570

[44] De, A., Mondal, N., Samanta, A., Journal of Physical Chemistry C, 122[25] 2018, 13617-13623. https://doi.org/10.1021/acs.jpcc.7b12813

[45] Chen, X., Peng, L., Huang, K., Shi, Z., Xie, R., Yang, W., Nano Research, 9[7] 2016, 1994-2006. https://doi.org/10.1007/s12274-016-1090-1

[46] Sun, S., Yuan, D., Xu, Y., Wang, A., Deng, Z., ACS Nano, 10[3] 2016, 3648-3657. https://doi.org/10.1021/acsnano.5b08193

[47] Ghosh, D., Ali, M.Y., Chaudhary, D.K., Bhattacharyya, S., Solar Energy Materials and Solar Cells, 185, 2018, 28-35. https://doi.org/10.1016/j.solmat.2018.05.002

[48] He, J., Vasenko, A.S., Long, R., Prezhdo, O.V., Journal of Physical Chemistry Letters, 9[8] 2018, 1872-1879. https://doi.org/10.1021/acs.jpclett.8b00446

[49] Wheeler, L.M., Sanehira, E.M., Marshall, A.R., Schulz, P., Suri, M., Anderson, N.C., Christians, J.A., Nordlund, D., Sokaras, D., Kroll, T., Harvey, S.P., Berry, J.J., Lin, L.Y., Luther, J.M., Journal of the American Chemical Society, 140[33] 2018, 10504-10513. https://doi.org/10.1021/jacs.8b04984

[50] Protesescu, L., Yakunin, S., Bodnarchuk, M.I., Krieg, F., Caputo, R., Hendon, C.H., Yang, R.X., Walsh, A., Kovalenko, M.V., Nano Letters, 15[6] 2015, 3692-3696. https://doi.org/10.1021/nl5048779

[51] Mondal, A., Aneesh, J., Kumar Ravi, V., Sharma, R., Mir, W.J., Beard, M.C., Nag, A., Adarsh, K.V., Physical Review B, 98[11] 2018, 115418. https://doi.org/10.1103/PhysRevB.98.115418

[52] Kumar, M., Pawar, V., Jha, P.A., Gupta, S.K., Sinha, A.S.K., Jha, P.K., Singh, P., Journal of Materials Science: Materials in Electronics, 30[6] 2019, 6071-6081. https://doi.org/10.1007/s10854-019-00908-x

[53] Amboy, J.M., Jeco, B.M.F.Y., Wang, H., Kubo, T., Kinoshita, T., Li-Kao, Z.J.,

Shoji, Y., Ahsan, N., Segawa, H., Okada, Y., Balbin, J.J.R. IEEE 7th World Conference on Photovoltaic Energy Conversion, 8548078, 2018, 467-471.

[54] Zhang, Y., Rong, M., Yan, X., Wang, X., Chen, Y., Li, X., Zhu, R., Langmuir, 34[32] 2018, 9507-9515. https://doi.org/10.1021/acs.langmuir.8b01650

[55] Ng, C.H., Lim, H.N., Hayase, S., Zainal, Z., Shafie, S., Lee, H.W., Huang, N.M., ACS Applied Energy Materials, 1[2] 2018, 692-699. https://doi.org/10.1021/acsaem.7b00103

[56] Lin, H.Y., Chen, C.Y., Hsu, B.W., Cheng, Y.L., Tsai, W.L., Huang, Y.C., Tsao, C.S., Lin, H.W., Advanced Functional Materials, 2019, 1905163. https://doi.org/10.1002/adfm.201905163

[57] Singh, A., Chouhan, A.S., Avasthi, S., Springer Proceedings in Physics, 215, 2019, 391-398. https://doi.org/10.1007/978-3-319-97604-4_60

[58] Wang, Y., Guan, X., Li, D., Cheng, H.C., Duan, X., Lin, Z., Duan, X., Nano Research, 10[4] 2017, 1223-1233. https://doi.org/10.1007/s12274-016-1317-1

[59] Zhang, P., Zhu, G., Shi, Y., Wang, Y., Zhang, J., Du, L., Ding, D., Journal of Physical Chemistry C, 122[48] 2018, 27148-27155. https://doi.org/10.1021/acs.jpcc.8b07237

[60] Fu, L., Zhang, Y., Chang, B., Li, B., Zhou, S., Zhang, L., Yin, L., Journal of Materials Chemistry A, 6[27] 2018, 13263-13270. https://doi.org/10.1039/C8TA02899K

[61] Liang, J., Liu, Z., Qiu, L., Hawash, Z., Meng, L., Wu, Z., Jiang, Y., Ono, L.K., Qi, Y., Advanced Energy Materials, 8[20] 2018, 1800504. https://doi.org/10.1002/aenm.201800504

[62] Yang, S., Wang, L., Gao, L., Cao, J., Han, Q., Yu, F., Kamata, Y., Zhang, C., Fan, M., Wei, G., Ma, T., ACS Applied Materials and Interfaces, 12[12] 2020, 13931-13940. https://doi.org/10.1021/acsami.9b23532

[63] He, J., Guo, M., Long, R., Journal of Physical Chemistry Letters, 9[11] 2018, 3021-3028. https://doi.org/10.1021/acs.jpclett.8b01266

[64] Yu, F.X., Zhang, Y., Xiong, Z.Y., Ma, X.J., Chen, P., Xiong, Z.H., Gao, C.H., Organic Electronics, 50, 2017, 480-484. https://doi.org/10.1016/j.orgel.2017.08.026

[65] Forde, A., Kilin, D., Journal of Physical Chemistry C, 121[37] 2017, 20113-20125. https://doi.org/10.1021/acs.jpcc.7b04961

[66] Gu, S., Zhu, P., Lin, R., Tang, M., Zhu, S., Zhu, J., Chinese Optics Letters, 15[9] 2017, 093501. https://doi.org/10.3788/COL201715.093501

[67] Zou, S., Liu, Y., Li, J., Liu, C., Feng, R., Jiang, F., Li, Y., Song, J., Zeng, H., Hong, M., Chen, X., Journal of the American Chemical Society, 139[33] 2017, 11443-11450. https://doi.org/10.1021/jacs.7b04000

[68] Yang, Z., Surrente, A., Galkowski, K., Miyata, A., Portugall, O., Sutton, R.J., Haghighirad, A.A., Snaith, H.J., Maude, D.K., Plochocka, P., Nicholas, R.J., ACS Energy Letters, 2[7] 2017, 1621-1627. https://doi.org/10.1021/acsenergylett.7b00416

[69] Liao, J.F., Li, W.G., Rao, H.S., Chen, B.X., Wang, X.D., Chen, H.Y., Kuang, D.B., Science China Materials, 60[4] 2017, 285-294. https://doi.org/10.1007/s40843-017-9014-9

[70] Nam, J.K., Chai, S.U., Cha, W., Choi, Y.J., Kim, W., Jung, M.S., Kwon, J., Kim, D., Park, J.H., Nano Letters, 17[3] 2017, 2028-2033. https://doi.org/10.1021/acs.nanolett.7b00050

[71] Guo, Y., Wang, Q., Saidi, W.A., Journal of Physical Chemistry C, 121[3] 2017, 1715-1722. https://doi.org/10.1021/acs.jpcc.6b11434

[72] Atourki, L., Vega, E., Mollar, M., Marí, B., Kirou, H., Bouabid, K., Ihlal, A., Journal of Alloys and Compounds, 702, 2017, 404-409. https://doi.org/10.1016/j.jallcom.2017.01.205

[73] Ramadan, A.J., Ralaiarisoa, M., Zu, F., Rochford, L.A., Wenger, B., Koch, N., Snaith, H.J., Chemistry of Materials, 32[1] 2020, 114-120. https://doi.org/10.1021/acs.chemmater.9b02639

[74] Beal, R.E., Slotcavage, D.J., Leijtens, T., Bowring, A.R., Belisle, R.A., Nguyen, W.H., Burkhard, G.F., Hoke, E.T., McGehee, M.D., Journal of Physical Chemistry Letters, 7[5] 2016, 746-751. https://doi.org/10.1021/acs.jpclett.6b00002

[75] Zhang, X., Lin, H., Huang, H., Reckmeier, C., Zhang, Y., Choy, W.C.H., Rogach, A.L., Nano Letters, 16[2] 2016, 1415-1420. https://doi.org/10.1021/acs.nanolett.5b04959

[76] Zhu, X., Lin, Y., San Martin, J., Sun, Y., Zhu, D., Yan, Y., Nature Communications, 10[1] 2019, 2843. https://doi.org/10.1038/s41467-019-10634-x

[77] Esch, M.P., Shu, Y., Levine, B.G., Journal of Physical Chemistry A, 123[13] 2019, 2661-2673. https://doi.org/10.1021/acs.jpca.9b00952

[78] Fernández-Delgado, N., Herrera, M., Delgado, F.J., Tavabi, A.H., Luysberg, M., Dunin-Borkowski, R.E., Juárez-Pérez, E.J., Hames, B.C., Mora-Sero, I., Suárez, I., Martinez-Pastor, J.P., Molina, S.I., Nanotechnology, 30[13] 2019, 135701. https://doi.org/10.1088/1361-6528/aafc85

[79] Wang, S.Q., Shen, S., Xue, X.X., He, Y., Xu, Z.W., Chen, K., Feng, Y., Applied Physics Express, 12[5] 2019, 051017. https://doi.org/10.7567/1882-0786/ab1a55

[80] Li, W., Vasenko, A.S., Tang, J., Prezhdo, O.V., Journal of Physical Chemistry Letters, 2019, 6219-6226. https://doi.org/10.1021/acs.jpclett.9b02553

[81] Chen, X., Wang, Z., Micron, 116, 2019, 73-79. https://doi.org/10.1016/j.micron.2018.09.010

[82] Pawar, V., Kumar, M., Jha, P.A., Gupta, S.K., Sinha, A.S.K., Jha, P.K., Singh, P., Journal of Thermal Analysis and Calorimetry, 139[5] 2020, 3073-3078. https://doi.org/10.1007/s10973-019-08676-w

[83] Geiregat, P., Maes, J., Chen, K., Drijvers, E., De Roo, J., Hodgkiss, J.M., Hens, Z., ACS Nano, 12[10] 2018, 10178-10188. https://doi.org/10.1021/acsnano.8b05092

[84] Zhou, H., Fan, L., He, G., Yuan, C., Wang, Y., Shi, S., Sui, N., Chen, B., Zhang, Y., Yao, Q., Zhao, J., Zhang, X., Yin, J., RSC Advances, 8[51] 2018, 29089-29095. https://doi.org/10.1039/C8RA04558E

[85] Yi, X., Zhang, Z., Chang, A., Mao, Y., Luan, Y., Lin, T., Wei, Y., Zhang, Y., Wang, F., Cao, S., Li, C., Wang, J., Advanced Energy Materials, 9[40] 2019, 1901726. https://doi.org/10.1002/aenm.201901726

[86] Aygüler, M.F., Weber, M.D., Puscher, B.M.D., Medina, D.D., Docampo, P., Costa, R.D., Journal of Physical Chemistry C, 119[21] 2015, 12047-12054. https://doi.org/10.1021/acs.jpcc.5b02959

[87] Leijtens, T., Prasanna, R., Bush, K.A., Eperon, G.E., Raiford, J.A., Gold-Parker, A., Wolf, E.J., Swifter, S.A., Boyd, C.C., Wang, H.P., Toney, M.F., Bent, S.F., McGehee, M.D., Sustainable Energy and Fuels, 2[11] 2018, 2450-2459. https://doi.org/10.1039/C8SE00314A

[88] Zheng, J., Deng, X., Zhou, X., Yu, M., Xia, Z., Chen, X., Huang, S., Journal of Materials Science - Materials in Electronics, 30[12] 2019, 11043-11053. https://doi.org/10.1007/s10854-019-01446-2

[89] Ndione, P.F., Li, Z., Zhu, K., Journal of Materials Chemistry C, 4[33] 2016, 7775-7782. https://doi.org/10.1039/C6TC02135B

[90] Wu, Y., Wang, P., Wang, S., Wang, Z., Cai, B., Zheng, X., Chen, Y., Yuan, N., Ding, J., Zhang, W.H., ChemSusChem, 11[5] 2018, 837-842. https://doi.org/10.1002/cssc.201702221

[91] Liu, G., Zheng, H., Xu, X., Zhu, L.Z., Zhang, X., Pan, X., Chemistry of Materials, 30[21] 2018, 7691-7698. https://doi.org/10.1021/acs.chemmater.8b02970

[92] Buizza, L.R.V., Crothers, T.W., Wang, Z., Patel, J.B., Milot, R.L., Snaith, H.J., Johnston, M.B., Herz, L.M., Advanced Functional Materials, 29[35] 2019, 1902656. https://doi.org/10.1002/adfm.201902656

[93] Bhatia, H., Steele, J.A., Martin, C., Keshavarz, M., Solis-Fernandez, G., Yuan, H., Fleury, G., Huang, H., Dovgaliuk, I., Chernyshov, D., Hendrix, J., Roeffaers, M.B.J., Hofkens, J., Debroye, E., Chemistry of Materials, 31[17] 2019, 6824-6832. https://doi.org/10.1021/acs.chemmater.9b01277

[94] Suzuki, A., Oku, T., Applied Surface Science, 483, 2019, 912-921. https://doi.org/10.1016/j.apsusc.2019.04.049

[95] McKenna, K.P., ACS Energy Letters, 3[11] 2018, 2663-2668. https://doi.org/10.1021/acsenergylett.8b01700

[96] Zhang, T., Li, H., Yang, P., Wei, J., Wang, F., Shen, H., Li, D., Li, F., Organic Electronics, 68, 2019, 76-84. https://doi.org/10.1016/j.orgel.2019.02.007

[97] Zhang, J., Wu, S., Liu, T., Zhu, Z., Jen, A.K.Y., Advanced Functional Materials, 2019, 1808833. https://doi.org/10.1002/adfm.201808833

[98] Whitcher, T.J., Zhu, J.X., Chi, X., Hu, H., Zhao, D., Asmara, T.C., Yu, X., Breese, M.B.H., Castro Neto, A.H., Lam, Y.M., Wee, A.T.S., Chia, E.E.M., Rusydi, A., Physical Review X, 8[2] 2018, 021034. https://doi.org/10.1103/PhysRevX.8.021034

[99] Lee, D.S., Yun, J.S., Kim, J., Soufiani, A.M., Chen, S., Cho, Y., Deng, X., Seidel, J., Lim, S., Huang, S., Ho-Baillie, A.W.Y., ACS Energy Letters, 3[3] 2018, 647-654. https://doi.org/10.1021/acsenergylett.8b00121

[100] Kato, M., Suzuki, A., Ohishi, Y., Tanaka, H., Oku, T., AIP Conference Proceedings, 1929, 2018, 020015.

[101] Poorkazem, K., Kelly, T.L., ACS Applied Energy Materials, 1[1] 2018, 181-190. https://doi.org/10.1021/acsaem.7b00065

[102] Ong, W.L., Elbaz, G., Doud, E.A., Kim, P., Paley, D., Roy, X., Malen, J.A., International Heat Transfer Conference, 2018, 6985-6992.

[103] Li, Q., Li, H., Shen, H., Wang, F., Zhao, F., Li, F., Zhang, X., Li, D., Jin, X., Sun, W., ACS Photonics, 4[10] 2017, 2504-2512. https://doi.org/10.1021/acsphotonics.7b00743

[104] Wang, Z., Lin, Q., Chmiel, F.P., Sakai, N., Herz, L.M., Snaith, H.J., Nature Energy, 2[9] 2017, 17135. https://doi.org/10.1038/nenergy.2017.135

[105] Indari, E.D., Wungu, T.D.K., Hidayat, R., Journal of Physics - Conference Series, 877[1] 2017, 012054. https://doi.org/10.1088/1742-6596/877/1/012054

[106] Yang, W.S., Park, B.W., Jung, E.H., Jeon, N.J., Kim, Y.C., Lee, D.U., Shin, S.S., Seo, J., Kim, E.K., Noh, J.H., Seok, S.I., Science, 356[6345] 2017, 1376-1379. https://doi.org/10.1126/science.aan2301

[107] Hills-Kimball, K., Nagaoka, Y., Cao, C., Chaykovsky, E., Chen, O., Journal of Materials Chemistry C, 5[23] 2017, 5680-5684. https://doi.org/10.1039/C7TC00598A

[108] Zhao, Z., Niu, Y., Zhao, Y., Song, Q., Xin, L., Lu, X., Acta Chimica Sinica, 74[8] 2016, 689-693. https://doi.org/10.6023/A16050245

[109] Zhumekenov, A.A., Saidaminov, M.I., Haque, M.A., Alarousu, E., Sarmah, S.P., Murali, B., Dursun, I., Miao, X.H., Abdelhady, A.L., Wu, T., Mohammed, O.F., Bakr, O.M., ACS Energy Letters, 1[1] 2016, 32-37. https://doi.org/10.1021/acsenergylett.6b00002

[110] Yun, J.S., Seidel, J., Kim, J., Soufiani, A.M., Huang, S., Lau, J., Jeon, N.J., Seok, S.I., Green, M.A., Ho-Baillie, A., Advanced Energy Materials, 6[13] 2016, 8. https://doi.org/10.1002/aenm.201600330

[111] Kirchartz, T., Philosophical Transactions of the Royal Society A, 377[2152] 2019, 20180286. https://doi.org/10.1098/rsta.2018.0286

[112] Hosokawa, H., Tamaki, R., Sawada, T., Okonogi, A., Sato, H., Ogomi, Y., Hayase, S., Okada, Y., Yano, T., Nature Communications, 10[1] 2019, 43. https://doi.org/10.1038/s41467-018-07655-3

[113] McGettrick, J.D., Hooper, K., Pockett, A., Baker, J., Troughton, J., Carnie, M., Watson, T., Materials Letters, 251, 2019, 98-101. https://doi.org/10.1016/j.matlet.2019.04.081

[114] Borghesi, C., Radicchi, E., Belpassi, L., Meggiolaro, D., De Angelis, F., Nunzi, F., Computational and Theoretical Chemistry, 1164, 2019, 112558. https://doi.org/10.1016/j.comptc.2019.112558

[115] Caselli, V.M., Fischer, M., Meggiolaro, D., Mosconi, E., De Angelis, F., Stranks, S.D., Baumann, A., Dyakonov, V., Hutter, E.M., Savenije, T.J., Journal of Physical Chemistry Letters, 10[17] 2019, 5128-5134. https://doi.org/10.1021/acs.jpclett.9b02160

[116] Piana, G.M., Bailey, C.G., Lagoudakis, P.G., Journal of Physical Chemistry C, 123[32] 2019, 19429-19436. https://doi.org/10.1021/acs.jpcc.9b06712

[117] Elmelund, T., Scheidt, R.A., Seger, B., Kamat, P.V., ACS Energy Letters, 4[8] 2019, 1961-1969. https://doi.org/10.1021/acsenergylett.9b01280

[118] Jiang, T., Shao, Z., Fang, H., Wang, W., Zhang, Q., Wu, D., Zhang, X., Jie, J., ACS Applied Materials and Interfaces, 11[27] 2019, 24367-24376. https://doi.org/10.1021/acsami.9b03474

[119] Shaban, A., Joodaki, M., Mehregan, S., Rangelow, I.W., Organic Electronics, 69, 2019, 106-113. https://doi.org/10.1016/j.orgel.2019.03.019

[120] Liu, D., Yang, C., Chen, P., Bates, M., Han, S., Askeland, P., Lunt, R.R., ACS Applied Energy Materials, 2[6] 2019, 3972-3978. https://doi.org/10.1021/acsaem.9b00270

[121] Crespo, C.T.,) Solar Energy Materials and Solar Cells, 195, 2019, 269-273. https://doi.org/10.1016/j.solmat.2019.03.023

[122] Motti, S.G., Crothers, T., Yang, R., Cao, Y., Li, R., Johnston, M.B., Wang, J., Herz, L.M., Nano Letters, 19[6] 2019, 3953-3960. https://doi.org/10.1021/acs.nanolett.9b01242

[123] Yang, F., Zhang, P., Kamarudin, M.A., Kapil, G., Ma, T., Hayase, S., Advanced Functional Materials, 28[46] 2018, 1804856. https://doi.org/10.1002/adfm.201804856

[124] Yamada, Y., Hoyano, M., Oto, K., Kanemitsu, Y., Physica Status Solidi B, 256[6] 2019, 1800545. https://doi.org/10.1002/pssb.201800545

[125] Mangalam, J., Rath, T., Weber, S., Kunert, B., Dimopoulos, T., Fian, A., Trimmel, G., Journal of Materials Science - Materials in Electronics, 30[10] 2019, 9602-9611. https://doi.org/10.1007/s10854-019-01294-0

[126] Papagiorgis, P., Manoli, A., Michael, S., Bernasconi, C., Bodnarchuk, M.I., Kovalenko, M.V., Othonos, A., Itskos, G., ACS Nano, 13[5] 2019, 5799-5809. https://doi.org/10.1021/acsnano.9b01398

[127] Gautier, R., Massuyeau, F., Galnon, G., Paris, M., Advanced Materials, 31[14]

2019, 1807383. https://doi.org/10.1002/adma.201807383

[128] Kim, B.G., Jang, W., Cho, J.S., Wang, D.H., Solar Energy Materials and Solar Cells, 192, 2019, 24-35. https://doi.org/10.1016/j.solmat.2018.12.010

[129] Laamari, M.E., Cheknane, A., Benghia, A., Hilal, H.S., Solar Energy, 182, 2019, 9-15. https://doi.org/10.1016/j.solener.2019.02.035

[130] Roose, B., Friend, R.H., Advanced Materials Interfaces, 6[5] 2019, 1801788. https://doi.org/10.1002/admi.201801788

[131] Liu, L., Xu, Y.Z., Tian, Y., Du, Y.W., Xin, C.G., Huang, W., An, S.C., Li, Y.L., Huang, Q., Hou, G.F., Zhao, Y., Zhang, X.D., Ding, Y., Journal of Synthetic Crystals, 48[3] 2019, 386-393.

[132] Nakada, K., Matsumoto, Y., Shimoi, Y., Yamada, K., Furukawa, Y., Molecules, 24[3] 2019, 626. https://doi.org/10.3390/molecules24030626

[133] Dhamaniya, B.P., Chhillar, P., Roose, B., Dutta, V., Pathak, S.K., ACS Applied Materials and Interfaces, 11[25] 2019, 22228-22239. https://doi.org/10.1021/acsami.9b00831

[134] García-Fernández, A., Moradi, Z., Bermúdez-García, J.M., Sánchez-Andújar, M., Gimeno, V.A., Castro-García, S., Senarís-Rodríguez, M.A., Mas-Marzá, E., Garcia-Belmonte, G., Fabregat-Santiago, F., Journal of Physical Chemistry C, 123[4] 2019, 2011-2018. https://doi.org/10.1021/acs.jpcc.8b03915

[135] Yadavalli, S.K., Dai, Z., Zhou, H., Zhou, Y., Padture, N.P., Acta Materialia, 187, 2020, 112-121. https://doi.org/10.1016/j.actamat.2020.01.040

[136] Xue, B., Bi, S., You, S., Zhou, J., Wu, G., Meng, R., Wang, B., Wang, J., Leng, X., Zhang, Y., Ma, X., Zhou, H., Chemistry - a European Journal, 25[4] 2019, 1076-1082.

[137] Aktas, E., Jiménez-López, J., Rodríguez-Seco, C., Pudi, R., Ortuño, M.A., López, N., Palomares, E., ChemPhysChem, 20[20] 2019, 2702-2711. https://doi.org/10.1002/cphc.201900068

[138] Chang, C., Zou, X., Cheng, J., Yang, Y., Yao, Y., Ling, T., Ren, H., Xiao, Z., Advances in Materials Science and Engineering, 2019, 2878060. https://doi.org/10.1155/2019/2878060

[139] Moreno-Romero, P.M., Corpus-Mendoza, A.N., Millán-Franco, M.A., Rodríguez-Castañeda, C.A., Torres-Herrera, D.M., Liu, F., Hu, H., Journal of Materials Science - Materials in Electronics, 30[18] 2019, 17491-17503.

https://doi.org/10.1007/s10854-019-02100-7

[140] Dhar, A., Khimani, M., Vekariya, R.L., Journal of Sol-Gel Science and Technology, 92[3] 2019, 548-553. https://doi.org/10.1007/s10971-019-05120-1

[141] Dhar, A., Dey, A., Maiti, P., Paul, P.K., Roy, S., Paul, S., Vekariya, R.L., Ionics, 24[4] 2018, 1227-1233. https://doi.org/10.1007/s11581-017-2256-x

[142] El-Mellouhi, F., Rashkeev, S.N., Marzouk, A., Kabalan, L., Belaidi, A., Merzougui, B., Tabet, N., Alharbi, F.H., Journal of Materials Chemistry C, 7[18] 2019, 5299-5306. https://doi.org/10.1039/C8TC06308G

[143] Li, T., Dunlap-Shohl, W.A., Reinheimer, E.W., Le Magueres, P., Mitzi, D.B., Chemical Science, 10[4] 2019, 1168-1175. https://doi.org/10.1039/C8SC03863E

[144] Gao, J., Liang, Q., Li, G., Ji, T., Liu, Y., Fan, M., Hao, Y., Liu, S.F., Wu, Y., Cui, Y., Journal of Materials Chemistry C, 7[27] 2019, 8357-8363. https://doi.org/10.1039/C9TC01309A

[145] Pazos-Outón, L.M., Xiao, T.P., Yablonovitch, E., IEEE 7th World Conference on Photovoltaic Energy Conversion, 8547204, 2018, 89-91.

[146] Li, W., Tang, J., Casanova, D., Prezhdo, O.V., ACS Energy Letters, 3[11] 2018, 2713-2720. https://doi.org/10.1021/acsenergylett.8b01608

[147] Yi, H.T., Irkhin, P., Joshi, P.P., Gartstein, Y.N., Zhu, X., Podzorov, V., Physical Review Applied, 10[5] 2018, 054016. https://doi.org/10.1103/PhysRevApplied.10.054016

[148] Mirershadi, S., Sattari, F., Saridaragh, M.M., Solar Energy Materials and Solar Cells, 186, 2018, 365-372. https://doi.org/10.1016/j.solmat.2018.07.008

[149] Zhang, X., Shen, J.X., Wang, W., Van De Walle, C.G., ACS Energy Letters, 3[10] 2018, 2329-2334. https://doi.org/10.1021/acsenergylett.8b01297

[150] Li, W., Zhou, L., Prezhdo, O.V., Akimov, A.V., ACS Energy Letters, 3[9] 2018, 2159-2166. https://doi.org/10.1021/acsenergylett.8b01226

[151] Hopper, T.R., Gorodetsky, A., Frost, J.M., Müller, C., Lovrincic, R., Bakulin, A.A., ACS Energy Letters, 3[9] 2018, 2199-2205. https://doi.org/10.1021/acsenergylett.8b01227

[152] Colella, S., Todaro, M., Masi, S., Listorti, A., Altamura, D., Caliandro, R., Giannini, C., Carignani, E., Geppi, M., Meggiolaro, D., Buscarino, G., De Angelis, F., Rizzo, A., ACS Energy Letters, 3[8] 2018, 1840-1847. https://doi.org/10.1021/acsenergylett.8b00944

[153] Sun, Q., Liu, X., Cao, J., Stantchev, R.I., Zhou, Y., Chen, X., Parrott, E.P.J., Lloyd-Hughes, J., Zhao, N., Pickwell-MacPherson, E., Journal of Physical Chemistry C, 122[30] 2018, 17552-17558. https://doi.org/10.1021/acs.jpcc.8b05695

[154] Kim, H.D., Ohkita, H., Japanese Journal of Applied Physics, 57[8] 2018, 08RE03. https://doi.org/10.7567/JJAP.57.08RE03

[155] Yerramilli, A.S., Chen, Y., Sanni, D., Asare, J., Theodore, N.D., Alford, T.L., Organic Electronics, 59, 2018, 107-112. https://doi.org/10.1016/j.orgel.2018.04.052

[156] Nurunnizar, A.A., Muslihun, Hidayat, R., Journal of Physics - Conference Series, 1057[1] 2018, 012007. https://doi.org/10.1088/1742-6596/1057/1/012007

[157] Abdi-Jalebi, M., Pazoki, M., Philippe, B., Dar, M.I., Alsari, M., Sadhanala, A., Divitini, G., Imani, R., Lilliu, S., Kullgren, J., Rensmo, H., Grätzel, M., Friend, R.H., ACS Nano, 12[7] 2018, 7301-7311. https://doi.org/10.1021/acsnano.8b03586

[158] Peng, W., Aranda, C., Bakr, O.M., Garcia-Belmonte, G., Bisquert, J., Guerrero, A., ACS Energy Letters, 3[7] 2018, 1477-1481. https://doi.org/10.1021/acsenergylett.8b00641

[159] Jacobsson, T.J., Svanström, S., Andrei, V., Rivett, J.P.H., Kornienko, N., Philippe, B., Cappel, U.B., Rensmo, H., Deschler, F., Boschloo, G., Journal of Physical Chemistry C, 122[25] 2018, 13548-13557. https://doi.org/10.1021/acs.jpcc.7b12464

[160] Zhu, F., Gentry, N.E., Men, L., White, M.A., Vela, J., Journal of Physical Chemistry C, 122[25] 2018, 14082-14090. https://doi.org/10.1021/acs.jpcc.8b01191

[161] Ono, L.K., Hawash, Z., Juarez-Perez, E.J., Qiu, L., Jiang, Y., Qi, Y., Journal of Physics D, 51[29] 2018, 294001. https://doi.org/10.1088/1361-6463/aacb6e

[162] Liu, L., Tang, Z., Xin, C., Zhu, S., An, S., Wang, N., Fan, L., Wei, C., Huang, Q., Hou, G., Zhao, Y., Ding, Y., Zhang, X., ACS Applied Energy Materials, 1[6] 2018, 2730-2739. https://doi.org/10.1021/acsaem.8b00400

[163] Kubicki, D.J., Prochowicz, D., Hofstetter, A., Zakeeruddin, S.M., Grätzel, M., Emsley, L., Journal of the American Chemical Society, 140[23] 2018, 7232-7238. https://doi.org/10.1021/jacs.8b03191

[164] Song, J., Xiao, Z., Chen, B., Prockish, S., Chen, X., Rajapitamahuni, A., Zhang, L., Huang, J., Hong, X., ACS Applied Materials and Interfaces, 10[22] 2018, 19218-

19225. https://doi.org/10.1021/acsami.8b03403

[165] Wang, Y., Lin, R., Zhu, P., Zheng, Q., Wang, Q., Li, D., Zhu, J., Nano Letters, 18[5] 2018, 2772-2779. https://doi.org/10.1021/acs.nanolett.7b04437

[166] Yu, W., Jiang, Y., Zhu, X., Luo, C., Jiang, K., Chen, L., Zhang, J., Chemical Physics Letters, 699, 2018, 93-98. https://doi.org/10.1016/j.cplett.2018.03.052

[167] Vega, E., Mollar, M., Marí, B., Journal of Alloys and Compounds, 739, 2018, 1059-1064. https://doi.org/10.1016/j.jallcom.2017.12.177

[168] Han, C., Wang, K., Zhu, X., Yu, H., Sun, X., Yang, Q., Hu, B., Journal of Physics D, 51[9] 2018, 095501. https://doi.org/10.1088/1361-6463/aaa7cd

[169] Ahmadian-Yazdi, M.R., Rahimzadeh, A., Chouqi, Z., Miao, Y., Eslamian, M., AIP Advances, 8[2] 2018, 025109. https://doi.org/10.1063/1.5019784

[170] Westbrook, R.J.E., Sanchez-Molina, D.I., Manuel Marin-Beloqui, D.J., Bronstein, D.H., Haque, D.S.A., Journal of Physical Chemistry C, 122[2] 2018, 1326-1332. https://doi.org/10.1021/acs.jpcc.7b09178

[171] Prasanna, R., Leijtens, T., Gold-Parker, A., Conings, B., Babayigit, A., Boyen, H.G., Toney, M.F., McGehee, M.D., IEEE 7th World Conference on Photovoltaic Energy Conversion, 8547344, 2018, 1718-1720.

[172] Li, C., Song, Z., Zhao, D., Xiao, C., Subedi, B., Shrestha, N., Junda, M.M., Wang, C., Jiang, C.S., Al-Jassim, M., Ellingson, R.J., Podraza, N.J., Zhu, K., Yan, Y., Advanced Energy Materials, 9[3] 2019, 1803135. https://doi.org/10.1002/aenm.201803135

[173] Mirershadi, S., Sattari, F., Sorkhabi, S.G., Shokri, A.M., Journal of Physical Chemistry C, 123[19] 2019, 12423-12428. https://doi.org/10.1021/acs.jpcc.9b02744

[174] Madjet, M.E., Berdiyorov, G.R., Ashhab, S., Computational Materials Science, 169, 2019, 109130. https://doi.org/10.1016/j.commatsci.2019.109130

[175] Chowdhury, T.H., Ferdaous, M.T., Wadi, M.A.A., Chelvanathan, P., Amin, N., Islam, A., Kamaruddin, N., Zin, M.I.M., Ruslan, M.H., Sopian, K.B., Akhtaruzzaman, M., Journal of Electronic Materials, 47[5] 2018, 3051-3058. https://doi.org/10.1007/s11664-018-6154-4

[176] Poindexter, J.R., Jensen, M.A., Morishige, A.E., Looney, E.E., Youssef, A., Correa-Baena, J.P., Wieghold, S., Rose, V., Lai, B., Cai, Z., Buonassisi, T., IEEE Journal of Photovoltaics, 8[1] 2018, 156-161.

https://doi.org/10.1109/JPHOTOV.2017.2775156

[177] Sun, M., Liang, C., Zhang, H., Ji, C., Sun, F., You, F., Jing, X., He, Z., Journal of Materials Chemistry A, 6[48] 2018, 24793-24804. https://doi.org/10.1039/C8TA07462C

[178] Xie, Z., Sun, S., Xie, X., Hou, R., Xu, W., Li, Y., Qin, G.G., Semiconductor Science and Technology, 33[1] 2018, 015011. https://doi.org/10.1088/1361-6641/aa9b24

[179] Šimenas, M., Banys, J., Tornau, E.E., Journal of Materials Chemistry C, 6[6] 2018, 1487-1494. https://doi.org/10.1039/C7TC05572B

[180] Busby, Y., Noël, C., Pescetelli, S., Agresti, A., Di Carlo, A., Pireaux, J.J., Houssiau, L., Proceedings of SPIE, 10724, 2018, 1072408.

[181] Svanström, S., Jacobsson, T.J., Sloboda, T., Giangrisostomi, E., Ovsyannikov, R., Rensmo, H., Cappel, U.B., Journal of Materials Chemistry A, 6[44] 2018, 22134-22144. https://doi.org/10.1039/C8TA05795H

[182] Kim, A., Son, B.H., Kim, H.S., Ahn, Y.H., Current Optics and Photonics, 2[6] 2018, 514-518.

[183] Pistor, P., Burwig, T., Brzuska, C., Weber, B., Fränzel, W., Journal of Materials Chemistry A, 6[24] 2018, 11496-11506. https://doi.org/10.1039/C8TA02775G

[184] Heiderhoff, R., Haeger, T., Pourdavoud, N., Hu, T., Al-Khafaji, M., Mayer, A., Chen, Y., Scheer, H.C., Riedl, T., Journal of Physical Chemistry C, 121[51] 2017, 28306-28311. https://doi.org/10.1021/acs.jpcc.7b11495

[185] Meyers, J.K., Kim, S., Hill, D.J., Cating, E.E.M., Williams, L.J., Kumbhar, A.S., McBride, J.R., Papanikolas, J.M., Cahoon, J.F., Nano Letters, 17[12] 2017, 7561-7568. https://doi.org/10.1021/acs.nanolett.7b03514

[186] Meggiolaro, D., Mosconi, E., De Angelis, F., ACS Energy Letters, 2[12] 2017, 2794-2798. https://doi.org/10.1021/acsenergylett.7b00955

[187] Love, J.A., Feuerstein, M., Wolff, C.M., Facchetti, A., Neher, D., ACS Applied Materials and Interfaces, 9[48] 2017, 42011-42019. https://doi.org/10.1021/acsami.7b10361

[188] Pazoki, M., Jacobsson, T.J., Cruz, S.H.T., Johansson, M.B., Imani, R., Kullgren, J., Hagfeldt, A., Johansson, E.M.J., Edvinsson, T., Boschloo, G., Journal of Physical Chemistry C, 121[47] 2017, 26180-26187. https://doi.org/10.1021/acs.jpcc.7b06775

[189] Li, C., Guerrero, A., Zhong, Y., Gräser, A., Luna, C.A.M., Köhler, J., Bisquert, J., Hildner, R., Huettner, S., Small, 13[42] 2017, 1701711. https://doi.org/10.1002/smll.201701711

[190] Iftiquar, S.M., Kim, J.S., Yi, J., Optik, 148, 2017, 54-62. https://doi.org/10.1016/j.ijleo.2017.08.141

[191] Burgués-Ceballos, I., Savva, A., Georgiou, E., Kapnisis, K., Papagiorgis, P., Mousikou, A., Itskos, G., Othonos, A., Choulis, S.A., AIP Advances, 7[11] 2017, 115304. https://doi.org/10.1063/1.5010261

[192] Wu, M.C., Lin, T.H., Chan, S.H., Su, W.F., Journal of the Taiwan Institute of Chemical Engineers, 80, 2017, 695-700. https://doi.org/10.1016/j.jtice.2017.09.004

[193] Bi, S., Zhang, X., Qin, L., Wang, R., Zhou, J., Leng, X., Qiu, X., Zhang, Y., Zhou, H., Tang, Z., Chemistry - a European Journal, 23[58] 2017, 14650-14657. https://doi.org/10.1002/chem.201703382

[194] Anusca, I., Balčiūnas, S., Gemeiner, P., Svirskas, Š., Sanlialp, M., Lackner, G., Fettkenhauer, C., Belovickis, J., Samulionis, V., Ivanov, M., Dkhil, B., Banys, J., Shvartsman, V.V., Lupascu, D.C., Advanced Energy Materials, 7[19] 2017, 1700600. https://doi.org/10.1002/aenm.201700600

[195] Park, B.W., Zhang, X., Johansson, E.M.J., Hagfeldt, A., Boschloo, G., Seok, S.I., Edvinsson, T., Nano Energy, 40, 2017, 596-606. https://doi.org/10.1016/j.nanoen.2017.08.055

[196] Olyaeefar, B., Ahmadi-Kandjani, S., Asgari, A., Physica E, 94, 2017, 118-122. https://doi.org/10.1016/j.physe.2017.07.018

[197] Kim, Y.C., Yang, T.Y., Jeon, N.J., Im, J., Jang, S., Shin, T.J., Shin, H.W., Kim, S., Lee, E., Kim, S., Noh, J.H., Seok, S.I., Seo, J., Energy and Environmental Science, 10[10] 2017, 2109-2116. https://doi.org/10.1039/C7EE01931A

[198] Yi, N., Wang, S., Duan, Z., Wang, K., Song, Q., Xiao, S., Advanced Materials, 29[34] 2017, 1701636. https://doi.org/10.1002/adma.201701636

[199] Cho, D.H., Choi, H.W., Molecular Crystals and Liquid Crystals, 654[1] 2017, 201-208. https://doi.org/10.1080/15421406.2017.1358045

[200] Ishioka, K., Barker, B.G., Yanagida, M., Shirai, Y., Miyano, K., Journal of Physical Chemistry Letters, 8[16] 2017, 3902-3907. https://doi.org/10.1021/acs.jpclett.7b01663

[201] Pazoki, M., Wolf, M.J., Edvinsson, T., Kullgren, J., Nano Energy, 38, 2017, 537-543. https://doi.org/10.1016/j.nanoen.2017.06.024

[202] Poindexter, J.R., Hoye, R.L.Z., Nienhaus, L., Kurchin, R.C., Morishige, A.E., Looney, E.E., Osherov, A., Correa-Baena, J.P., Lai, B., Bulović, V., Stevanović, V., Bawendi, M.G., Buonassisi, T., ACS Nano, 11[7] 2017, 7101-7109. https://doi.org/10.1021/acsnano.7b02734

[203] Yang, H., Zhang, J., Zhang, C., Chang, J., Lin, Z., Chen, D., Xi, H., Hao, Y., Materials, 10[7] 2017, 837. https://doi.org/10.3390/ma10070837

[204] Vogel, D.J., Kryjevski, A., Inerbaev, T., Kilin, D.S., Journal of Physical Chemistry Letters, 8[13] 2017, 3032-3039. https://doi.org/10.1021/acs.jpclett.6b03048

[205] Ma, D., Fu, Y., Dang, L., Zhai, J., Guzei, I.A., Jin, S., Nano Research, 10[6] 2017, 2117-2129. https://doi.org/10.1007/s12274-016-1401-6

[206] Aamir, M., Shah, Z.H., Sher, M., Iqbal, A., Revaprasadu, N., Malik, M.A., Akhtar, J., Materials Science in Semiconductor Processing, 63, 2017, 6-11. https://doi.org/10.1016/j.mssp.2017.01.001

[207] Idowu, M.A., Yau, S.H., Varnavski, O., Goodson, T., ACS Photonics, 4[5] 2017, 1195-1206. https://doi.org/10.1021/acsphotonics.7b00095

[208] Plugaru, N., Nemnes, G.A., Filip, L., Pintilie, I., Pintilie, L., Butler, K.T., Manolescu, A., Journal of Physical Chemistry C, 121[17] 2017, 9096-9109. https://doi.org/10.1021/acs.jpcc.7b00399

[209] Game, O.S., Buchsbaum, G.J., Zhou, Y., Padture, N.P., Kingon, A.I., Advanced Functional Materials, 27[16] 2017, 1606584. https://doi.org/10.1002/adfm.201606584

[210] Baker, J.A., Mouhamad, Y., Hooper, K.E.A., Burkitt, D., Geoghegan, M., Watson, T.M., IET Renewable Power Generation, 11[5] 2017, 546-549. https://doi.org/10.1049/iet-rpg.2016.0683

[211] Stevenson, J., Sorenson, B., Subramaniam, V.H., Raiford, J., Khlyabich, P.P., Loo, Y.L., Clancy, P., Chemistry of Materials, 29[6] 2017, 2435-2444. https://doi.org/10.1021/acs.chemmater.6b04327

[212] Kovalenko, A., Pospisil, J., Zmeskal, O., Krajcovic, J., Weiter, M., Physica Status Solidi - Rapid Research Letters, 11[3] 2017, 1600418. https://doi.org/10.1002/pssr.201600418

[213] Chen, S., Wen, X., Yun, J.S., Huang, S., Green, M., Jeon, N.J., Yang, W.S., Noh,

J.H., Seo, J., Seok, S.I., Ho-Baillie, A., ACS Applied Materials and Interfaces, 9[7] 2017, 6072-6078. https://doi.org/10.1021/acsami.6b15504

[214] Costa, J.C.S., Azevedo, J., Santos, L.M.N.B.F., Mendes, A., Journal of Physical Chemistry C, 121[4] 2017, 2080-2087. https://doi.org/10.1021/acs.jpcc.6b11625

[215] Liu, M., Endo, M., Shimazaki, A., Wakamiya, A., Tachibana, Y., Journal of Photopolymer Science and Technology, 30[5] 2017, 577-582. https://doi.org/10.2494/photopolymer.30.577

[216] Hendriks, K.H., Van Franeker, J.J., Bruijnaers, B.J., Anta, J.A., Wienk, M.M., Janssen, R.A.J., Journal of Materials Chemistry A, 5[5] 2017, 2346-2354. https://doi.org/10.1039/C6TA09125C

[217] Choe, G., Kang, J., Ryu, I., Song, S.W., Kim, H.M., Yim, S., Solar Energy, 155, 2017, 1148-1156. https://doi.org/10.1016/j.solener.2017.07.065

[218] Belisle, R.A., Nguyen, W.H., Bowring, A.R., Calado, P., Li, X., Irvine, S.J.C., McGehee, M.D., Barnes, P.R.F., O'Regan, B.C., Energy and Environmental Science, 10[1] 2017, 192-204. https://doi.org/10.1039/C6EE02914K

[219] Du, T., Burgess, C.H., Kim, J., Zhang, J., Durrant, J.R., McLachlan, M.A., Sustainable Energy and Fuels, 1[1] 2017, 119-126. https://doi.org/10.1039/C6SE00029K

[220] Cotella, G., Baker, J., Worsley, D., De Rossi, F., Pleydell-Pearce, C., Carnie, M., Watson, T., Solar Energy Materials and Solar Cells, 159, 2017, 362-369. https://doi.org/10.1016/j.solmat.2016.09.013

[221] Chen, G., Zhang, F., Liu, M., Song, J., Lian, J., Zeng, P., Yip, H.L., Yang, W., Zhang, B., Cao, Y., Journal of Materials Chemistry A, 5[34] 2017, 17943-17953. https://doi.org/10.1039/C7TA04995A

[222] Zhu, H., Trinh, M.T., Wang, J., Fu, Y., Joshi, P.P., Miyata, K., Jin, S., Zhu, X.Y., Advanced Materials, 29[1] 2017, 1603072. https://doi.org/10.1002/adma.201603072

[223] Sampson, M.D., Park, J.S., Schaller, R.D., Chan, M.K.Y., Martinson, A.B.F., Journal of Materials Chemistry A, 5[7] 2017, 3578-3588. https://doi.org/10.1039/C6TA09745F

[224] Ščajev, P., Aleksiejunas, R., Miasojedovas, S., Nargelas, S., Inoue, M., Qin, C., Matsushima, T., Adachi, C., Juršenas, S., Journal of Physical Chemistry C, 121[39] 2017, 21600-21609. https://doi.org/10.1021/acs.jpcc.7b04179

[225] Šimenas, M., Balčiunas, S., Mączka, M., Banys, J., Tornau, E.E., Journal of Physical Chemistry Letters, 8[19] 2017, 4906-4911. https://doi.org/10.1021/acs.jpclett.7b02239

[226] Khlyabich, P.P., Loo, Y.L., Chemistry of Materials, 28[24] 2016, 9041-9048. https://doi.org/10.1021/acs.chemmater.6b04020

[227] Richter, J.M., Abdi-Jalebi, M., Sadhanala, A., Tabachnyk, M., Rivett, J.P.H., Pazos-Outón, L.M., Gödel, K.C., Price, M., Deschler, F., Friend, R.H., Nature Communications, 7, 2016, 13941. https://doi.org/10.1038/ncomms13941

[228] Lilliu, S., Dane, T.G., Alsari, M., Griffin, J., Barrows, A.T., Dahlem, M.S., Friend, R.H., Lidzey, D.G., Macdonald, J.E., Advanced Functional Materials, 26[45] 2016, 8221-8230. https://doi.org/10.1002/adfm.201603446

[229] Sanchez, R.S., Mas-Marza, E., Solar Energy Materials and Solar Cells, 158, 2016, 189-194. https://doi.org/10.1016/j.solmat.2016.03.024

[230] Motti, S.G., Gandini, M., Barker, A.J., Ball, J.M., Srimath Kandada, A.R., Petrozza, A., ACS Energy Letters, 1[4] 2016, 726-730. https://doi.org/10.1021/acsenergylett.6b00355

[231] Huang, Z., Duan, X., Zhang, Y., Hu, X., Tan, L., Chen, Y., Solar Energy Materials and Solar Cells, 155, 2016, 166-175. https://doi.org/10.1016/j.solmat.2016.06.011

[232] Dar, M.I., Jacopin, G., Meloni, S., Mattoni, A., Arora, N., Boziki, A., Zakeeruddin, S.M., Rothlisberger, U., Grätzel, M., Science Advances, 2[10] 2016, 1601156. https://doi.org/10.1126/sciadv.1601156

[233] Namkoong, G., Jeong, H.J., Mamun, A., Byun, H., Demuth, D., Jeong, M.S., Solar Energy Materials and Solar Cells, 155, 2016, 134-140. https://doi.org/10.1016/j.solmat.2016.06.008

[234] Kim, D., Kim, G.Y., Ko, C., Pae, S.R., Lee, Y.S., Gunawan, O., Ogletree, D.F., Jo, W., Shin, B., Journal of Physical Chemistry C, 120[38] 2016, 21330-21335. https://doi.org/10.1021/acs.jpcc.6b08744

[235] Rahimnejad, S., Kovalenko, A., Forés, S.M., Aranda, C., Guerrero, A., ChemPhysChem, 2016, 2795-2798. https://doi.org/10.1002/cphc.201600575

[236] Matsuda, T., Tateishi, Y., Yamashita, K., Kitamura, M., Kita, T., Journal of the Society of Materials Science, Japan, 65[9] 2016, 642-646. https://doi.org/10.2472/jsms.65.642

[237] Brunetti, B., Cavallo, C., Ciccioli, A., Gigli, G., Latini, A., Scientific Reports, 6,

2016, 31896. https://doi.org/10.1038/srep31896

[238] Chang, A.Y., Cho, Y.J., Chen, K.C., Chen, C.W., Kinaci, A., Diroll, B.T., Wagner, M.J., Chan, M.K.Y., Lin, H.W., Schaller, R.D., Advanced Energy Materials, 6[15] 2016, 1600422. https://doi.org/10.1002/aenm.201600422

[239] Zhang, T., Guo, N., Li, G., Qian, X., Zhao, Y., Nano Energy, 26, 2016, 50-56. https://doi.org/10.1016/j.nanoen.2016.05.003

[240] Liang, D., Peng, Y., Fu, Y., Shearer, M.J., Zhang, J., Zhai, J., Zhang, Y., Hamers, R.J., Andrew, T.L., Jin, S., ACS Nano, 10[7] 2016, 6897-6904. https://doi.org/10.1021/acsnano.6b02683

[241] Nagabhushana, G.P., Shivaramaiah, R., Navrotsky, A., Proceedings of the National Academy of Sciences, 113[28] 2016, 7717-7721. https://doi.org/10.1073/pnas.1607850113

[242] Jiang, Y., Wen, X., Benda, A., Sheng, R., Ho-Baillie, A.W.Y., Huang, S., Huang, F., Cheng, Y.B., Green, M.A., Solar Energy Materials and Solar Cells, 151, 2016, 102-112. https://doi.org/10.1016/j.solmat.2016.03.002

[243] Guerrero, A., Garcia-Belmonte, G., Mora-Sero, I., Bisquert, J., Kang, Y.S., Jacobsson, T.J., Correa-Baena, J.P., Hagfeldt, A., Journal of Physical Chemistry C, 120[15] 2016, 8023-8032. https://doi.org/10.1021/acs.jpcc.6b01728

[244] Song, D., Wei, D., Cui, P., Li, M., Duan, Z., Wang, T., Ji, J., Li, Y., Mbengue, J.M., Li, Y., He, Y., Trevor, M., Park, N.G., Journal of Materials Chemistry A, 4[16] 2016, 6091-6097. https://doi.org/10.1039/C6TA00577B

[245] Jaffe, A., Lin, Y., Beavers, C.M., Voss, J., Mao, W.L., Karunadasa, H.I., ACS Central Science, 2[4] 2016, 201-209. https://doi.org/10.1021/acscentsci.6b00055

[246] Kong, W., Rahimi-Iman, A., Bi, G., Dai, X., Wu, H., Journal of Physical Chemistry C, 120[14] 2016, 7606-7611. https://doi.org/10.1021/acs.jpcc.6b00496

[247] Jaramillo-Quintero, O.A., Solís De La Fuente, M., Sanchez, R.S., Recalde, I.B., Juarez-Perez, E.J., Rincón, M.E., Mora-Seró, I., Nanoscale, 8[12] 2016, 6271-6277. https://doi.org/10.1039/C5NR06692A

[248] Kim, Y.C., Jeon, N.J., Noh, J.H., Yang, W.S., Seo, J., Yun, J.S., Ho-Baillie, A., Huang, S., Green, M.A., Seidel, J., Ahn, T.K., Seok, S.I., Advanced Energy Materials, 6[4] 2016, 1502104. https://doi.org/10.1002/aenm.201502104

[249] Miyano, K., Tripathi, N., Yanagida, M., Shirai, Y., Accounts of Chemical Research, 49[2] 2016, 303-310. https://doi.org/10.1021/acs.accounts.5b00436

[250] Wei, D., Wang, T., Ji, J., Li, M., Cui, P., Li, Y., Li, G., Mbengue, J.M., Song, D., Journal of Materials Chemistry A, 4[5] 2016, 1991-1998. https://doi.org/10.1039/C5TA08622A

[251] Ebe, H., Araki, H., Japanese Journal of Applied Physics, 55[2] 2016, 02BF11. https://doi.org/10.7567/JJAP.55.02BF11

[252] Qaid, S.M.H., Al Sobaie, M.S., Majeed Khan, M.A., Bedja, I.M., Alharbi, F.H., Nazeeruddin, M.K., Aldwayyan, A.S., Materials Letters, 164, 2016, 498-501. https://doi.org/10.1016/j.matlet.2015.10.135

[253] Ye, T., Fu, W., Wu, J., Yu, Z., Jin, X., Chen, H., Li, H., Journal of Materials Chemistry A, 4[4] 2016, 1214-1217. https://doi.org/10.1039/C5TA10155G

[254] Guerrero, A., You, J., Aranda, C., Kang, Y.S., Garcia-Belmonte, G., Zhou, H., Bisquert, J., Yang, Y., ACS Nano, 10[1] 2016, 218-224. https://doi.org/10.1021/acsnano.5b03687

[255] Almora, O., Guerrero, A., Garcia-Belmonte, G., Applied Physics Letters, 108[4] 2016, 043903. https://doi.org/10.1063/1.4941033

[256] Peng, W., Anand, B., Liu, L., Sampat, S., Bearden, B.E., Malko, A.V., Chabal, Y.J., Nanoscale, 8[3] 2016, 1627-1634. https://doi.org/10.1039/C5NR06222E

[257] Andalibi, S., Rostami, A., Darvish, G., Moravvej-Farshi, M.K., Optical and Quantum Electronics, 48[4] 2016, 258. https://doi.org/10.1007/s11082-016-0525-y

[258] Jishi, R.A., AIMS Materials Science, 3[1] 2016, 149-159. https://doi.org/10.3934/matersci.2016.1.149

[259] Troughton, J., Carnie, M.J., Davies, M.L., Charbonneau, C., Jewell, E.H., Worsley, D.A., Watson, T.M., Journal of Materials Chemistry A, 4[9] 2016, 3471-3476. https://doi.org/10.1039/C5TA09431C

[260] Wang, B., Wong, K.Y., Yang, S., Chen, T., Journal of Materials Chemistry A, 4[10] 2016, 3806-3812. https://doi.org/10.1039/C5TA09249C

[261] Conings, B., Babayigit, A., Klug, M.T., Bai, S., Gauquelin, N., Sakai, N., Wang, J.T.W., Verbeeck, J., Boyen, H.G., Snaith, H.J., Advanced Materials, 28[48] 2016, 10701-10709. https://doi.org/10.1002/adma.201603747

[262] Zhang, Y., Huang, F., Mi, Q., Chemistry Letters, 45[8] 2016, 1030-1032. https://doi.org/10.1246/cl.160419

[263] Murali, B., Saidaminov, M.I., Abdelhady, A.L., Peng, W., Liu, J., Pan, J., Bakr, O.M., Mohammed, O.F., Journal of Materials Chemistry C, 4[13] 2016, 2545-

2552. https://doi.org/10.1039/C6TC00610H

[264] Stumpp, M., Ruess, R., Horn, J., Tinz, J., Richter, C., Schlettwein, D., Physica Status Solidi A, 213[1] 2016, 38-45. https://doi.org/10.1002/pssa.201532527

[265] Guse, J.A., Soufiani, A.M., Jiang, L., Kim, J., Cheng, Y.B., Schmidt, T.W., Ho-Baillie, A., McCamey, D.R., Physical Chemistry Chemical Physics, 18[17] 2016, 12043-12049. https://doi.org/10.1039/C5CP07360J

[266] Kong, W., Ding, T., Bi, G., Wu, H., Physical Chemistry Chemical Physics, 18[18] 2016, 12626-12632. https://doi.org/10.1039/C6CP00325G

[267] Soufiani, A.M., Huang, F., Reece, P., Sheng, R., Ho-Baillie, A., Green, M.A., Applied Physics Letters, 107[23] 2015, 231902. https://doi.org/10.1063/1.4936418

[268] Jiang, M., Wu, J., Lan, F., Tao, Q., Gao, D., Li, G., Journal of Materials Chemistry A, 3[3] 2015, 963-967. https://doi.org/10.1039/C4TA05373G

[269] Sestu, N., Cadelano, M., Sarritzu, V., Chen, F., Marongiu, D., Piras, R., Mainas, M., Quochi, F., Saba, M., Mura, A., Bongiovanni, G., Journal of Physical Chemistry Letters, 6[22] 2015, 4566-4572. https://doi.org/10.1021/acs.jpclett.5b02099

[270] Hossain, M.I., Alharbi, F.H., Tabet, N., Solar Energy, 120, 2015, 370-380. https://doi.org/10.1016/j.solener.2015.07.040

[271] Song, D., Cui, P., Wang, T., Wei, D., Li, M., Cao, F., Yue, X., Fu, P., Li, Y., He, Y., Jiang, B., Trevor, M., Journal of Physical Chemistry C, 119[40] 2015, 22812-22819. https://doi.org/10.1021/acs.jpcc.5b06859

[272] Borchert, J., Boht, H., Fränzel, W., Csuk, R., Scheer, R., Pistor, P., Journal of Materials Chemistry A, 3[39] 2015, 19842-19849. https://doi.org/10.1039/C5TA04944J

[273] Wang, M., Shi, C., Zhang, J., Wu, N., Ying, C., Journal of Solid State Chemistry, 231, 2015, 19026, 20-24. https://doi.org/10.1016/j.jssc.2015.08.002

[274] Kim, G.Y., Oh, S.H., Nguyen, B.P., Jo, W., Kim, B.J., Lee, D.G., Jung, H.S., Journal of Physical Chemistry Letters, 6[12] 2015, 2355-2362. https://doi.org/10.1021/acs.jpclett.5b00967

[275] Almora, O., Zarazua, I., Mas-Marza, E., Mora-Sero, I., Bisquert, J., Garcia-Belmonte, G., Journal of Physical Chemistry Letters, 6[9] 2015, 1645-1652. https://doi.org/10.1021/acs.jpclett.5b00480

[276] Almond, D.P., Bowen, C.R., Journal of Physical Chemistry Letters, 6[9] 2015,

1736-1740. https://doi.org/10.1021/acs.jpclett.5b00620

[277] Fu, Y., Meng, F., Rowley, M.B., Thompson, B.J., Shearer, M.J., Ma, D., Hamers, R.J., Wright, J.C., Jin, S., Journal of the American Chemical Society, 137[17] 2015, 5810-5818. https://doi.org/10.1021/jacs.5b02651

[278] Wang, Q.K., Wang, R.B., Shen, P.F., Li, C., Li, Y.Q., Liu, L.J., Duhm, S., Tang, J.X., Advanced Materials Interfaces, 2[3] 2015, 1400528. https://doi.org/10.1002/admi.201400528

[279] Chen, Y.S., Manser, J.S., Kamat, P.V., Journal of the American Chemical Society, 137[2] 2015, 974-981. https://doi.org/10.1021/ja511739y

[280] Ito, S., Tanaka, S., Nishino, H., Journal of Physical Chemistry Letters, 6[5] 2015, 881-886. https://doi.org/10.1021/acs.jpclett.5b00122

[281] Hossain, M., Alharbi, F., Tabet, N., Proceedings of the TMS Middle East - Mediterranean Materials Congress on Energy and Infrastructure Systems, MEMA 2015, 339-342. https://doi.org/10.1002/9781119090427.ch35

[282] Qin, P., Tanaka, S., Ito, S., Tetreault, N., Manabe, K., Nishino, H., Nazeeruddin, M.K., Grätzel, M., Nature Communications, 5, 2014, 3834. https://doi.org/10.1038/ncomms4834

[283] Christians, J.A., Fung, R.C.M., Kamat, P.V., Journal of the American Chemical Society, 136[2] 2014, 758-764. https://doi.org/10.1021/ja411014k

[284] Bryant, D., Greenwood, P., Troughton, J., Wijdekop, M., Carnie, M., Davies, M., Wojciechowski, K., Snaith, H.J., Watson, T., Worsley, D., Advanced Materials, 26[44] 2014, 7499-7504. https://doi.org/10.1002/adma.201403939

[285] Coll, M., Gomez, A., Mas-Marza, E., Almora, O., Garcia-Belmonte, G., Campoy-Quiles, M., Bisquert, J., Journal of Physical Chemistry Letters, 6[8] 2015, 1408-1413. https://doi.org/10.1021/acs.jpclett.5b00502

[286] Juarez-Perez, E.J., Sanchez, R.S., Badia, L., Garcia-Belmonte, G., Kang, Y.S., Mora-Sero, I., Bisquert, J., Journal of Physical Chemistry Letters, 5[13] 2014, 2390-2394. https://doi.org/10.1021/jz5011169

[287] Leijtens, T., Hoke, E.T., Grancini, G., Slotcavage, D.J., Eperon, G.E., Ball, J.M., De Bastiani, M., Bowring, A.R., Martino, N., Wojciechowski, K., McGehee, M.D., Snaith, H.J., Petrozza, A., Advanced Energy Materials, 5[20] 2015, 1500962. https://doi.org/10.1002/aenm.201500962

[288] Raoui, Y., Ez-Zahraouy, H., Tahiri, N., El Bounagui, O., Ahmad, S., Kazim, S.,

Solar Energy, 193, 2019, 948-955. https://doi.org/10.1016/j.solener.2019.10.009

[289] Pockett, A., Eperon, G.E., Peltola, T., Snaith, H.J., Walker, A., Peter, L.M., Cameron, P.J., Journal of Physical Chemistry C, 119[7] 2015, 3456-3465. https://doi.org/10.1021/jp510837q

[290] Stamplecoskie, K.G., Manser, J.S., Kamat, P.V., Energy and Environmental Science, 8[1] 2015, 208-215. https://doi.org/10.1039/C4EE02988G

[291] Grancini, G., Srimath Kandada, A.R., Frost, J.M., Barker, A.J., De Bastiani, M., Gandini, M., Marras, S., Lanzani, G., Walsh, A., Petrozza, A., Nature Photonics, 9[10] 2015, 695-701. https://doi.org/10.1038/nphoton.2015.151

[292] Wang, D., Liu, Z., Zhou, Z., Zhu, H., Zhou, Y., Huang, C., Wang, Z., Xu, H., Jin, Y., Fan, B., Pang, S., Cui, G., Chemistry of Materials, 26[24] 2014, 7145-7150. https://doi.org/10.1021/cm5037869

[293] Dharani, S., Dewi, H.A., Prabhakar, R.R., Baikie, T., Shi, C., Yonghua, D., Mathews, N., Boix, P.P., Mhaisalkar, S.G., Nanoscale, 6[22] 2014, 13854-13860. https://doi.org/10.1039/C4NR04007D

[294] Tidhar, Y., Edri, E., Weissman, H., Zohar, D., Hodes, G., Cahen, D., Rybtchinski, B., Kirmayer, S., Journal of the American Chemical Society, 136[38] 2014, 13249-13256. https://doi.org/10.1021/ja505556s

[295] Grote, C., Ehrlich, B., Berger, R.F., Physical Review B, 90[20] 2014, 205202. https://doi.org/10.1103/PhysRevB.90.205202

[296] Guerrero, A., Juarez-Perez, E.J., Bisquert, J., Mora-Sero, I., Garcia-Belmonte, G., Applied Physics Letters, 105[13] 2014, 4896779. https://doi.org/10.1063/1.4896779

[297] Park, B.W., Philippe, B., Gustafsson, T., Sveinbjörnsson, K., Hagfeldt, A., Johansson, E.M.J., Boschloo, G., Chemistry of Materials, 26[15] 2014, 4466-4471. https://doi.org/10.1021/cm501541p

[298] Aharon, S., Cohen, B.E., Etgar, L., Journal of Physical Chemistry C, 118[30] 2014, 17160-17165. https://doi.org/10.1021/jp5023407

[299] Wang, Q., Yun, J.H., Zhang, M., Chen, H., Chen, Z.G., Wang, L., Journal of Materials Chemistry A, 2[27] 2014, 10355-10358. https://doi.org/10.1039/c4ta01105h

[300] Juarez-Perez, E.J., Wußler, M., Fabregat-Santiago, F., Lakus-Wollny, K., Mankel, E., Mayer, T., Jaegermann, W., Mora-Sero, I., Journal of Physical Chemistry

Letters, 5[4] 2014, 680-685. https://doi.org/10.1021/jz500059v

[301] Noel, N.K., Abate, A., Stranks, S.D., Parrott, E.S., Burlakov, V.M., Goriely, A., Snaith, H.J., ACS Nano, 8[10] 2014, 9815-9821. https://doi.org/10.1021/nn5036476

[302] Cohen, B.E., Gamliel, S., Etgar, L., APL Materials, 2[8] 2014, 081502. https://doi.org/10.1063/1.4885548

[303] Jin, Y., Chumanov, G., Chemistry Letters, 43[11] 2014, 1722-1724. https://doi.org/10.1246/cl.140644

[304] Pistor, P., Borchert, J., Fränzel, W., Csuk, R., Scheer, R., Journal of Physical Chemistry Letters, 5[19] 2014, 3308-3312. https://doi.org/10.1021/jz5017312

[305] Wehrenfennig, C., Liu, M., Snaith, H.J., Johnston, M.B., Herz, L.M., APL Materials, 2[8] 2014, 081513. https://doi.org/10.1063/1.4891595

[306] Laban, W.A., Etgar, L., Energy and Environmental Science, 6[11] 2013, 3249-3253. https://doi.org/10.1039/c3ee42282h

[307] Sanchez, R.S., Gonzalez-Pedro, V., Lee, J.W., Park, N.G., Kang, Y.S., Mora-Sero, I., Bisquert, J., Journal of Physical Chemistry Letters, 5[13] 2014, 2357-2363. https://doi.org/10.1021/jz5011187

[308] Chen, H.W., Sakai, N., Ikegami, M., Miyasaka, T., Journal of Physical Chemistry Letters, 6[1] 2015, 164-169. https://doi.org/10.1021/jz502429u

[309] Bryant, D., Wheeler, S., O'Regan, B.C., Watson, T., Barnes, P.R.F., Worsley, D., Durrant, J., Journal of Physical Chemistry Letters, 6[16] 2015, 3190-3194. https://doi.org/10.1021/acs.jpclett.5b01381

[310] Schoonman, J., Chemical Physics Letters, 619, 2015, 193-195. https://doi.org/10.1016/j.cplett.2014.11.063

[311] Gao, H., Bao, C., Li, F., Yu, T., Yang, J., Zhu, W., Zhou, X., Fu, G., Zou, Z., ACS Applied Materials and Interfaces, 7[17] 2015, 9110-9117. https://doi.org/10.1021/acsami.5b00895

[312] Philippe, B., Park, B.W., Lindblad, R., Oscarsson, J., Ahmadi, S., Johansson, E.M.J., Rensmo, H., Chemistry of Materials, 27[5] 2015, 1720-1731. https://doi.org/10.1021/acs.chemmater.5b00348

[313] Eperon, G.E., Habisreutinger, S.N., Leijtens, T., Bruijnaers, B.J., Van Franeker, J.J., Dequilettes, D.W., Pathak, S., Sutton, R.J., Grancini, G., Ginger, D.S., Janssen, R.A.J., Petrozza, A., Snaith, H.J., ACS Nano, 9[9] 2015, 9380-9393.

https://doi.org/10.1021/acsnano.5b03626

[314] Zhang, L., Ju, M.G., Liang, W., Physical Chemistry Chemical Physics, 18[33] 2016, 23174-23183. https://doi.org/10.1039/C6CP01994C

[315] Chaudhary, B., Kulkarni, A., Jena, A.K., Ikegami, M., Udagawa, Y., Kunugita, H., Ema, K., Miyasaka, T., ChemSusChem, 10[11] 2017, 2473-2479. https://doi.org/10.1002/cssc.201700271

[316] Wu, F., Zhu, L., Solar Energy Materials and Solar Cells, 167, 2017, 1-6. https://doi.org/10.1016/j.solmat.2017.03.030

[317] Jong, U.G., Yu, C.J., Ri, G.C., McMahon, A.P., Harrison, N.M., Barnes, P.R.F., Walsh, A., Journal of Materials Chemistry A, 6[3] 2018, 1067-1074. https://doi.org/10.1039/C7TA09112E

[318] Barbé, J., Kumar, V., Newman, M.J., Lee, H.K.H., Jain, S.M., Chen, H., Charbonneau, C., Rodenburg, C., Tsoi, W.C., Sustainable Energy and Fuels, 2[4] 2018, 905-914. https://doi.org/10.1039/C7SE00545H

[319] Turkevych, I., Kazaoui, S., Ito, E., Urano, T., Yamada, K., Tomiyasu, H., Yamagishi, H., Kondo, M., Aramaki, S., ChemSusChem, 10[19] 2017, 3754-3759. https://doi.org/10.1002/cssc.201700980

[320] Slavney, A.H., Smaha, R.W., Smith, I.C., Jaffe, A., Umeyama, D., Karunadasa, H.I., Inorganic Chemistry, 56[1] 2017, 46-55. https://doi.org/10.1021/acs.inorgchem.6b01336

[321] Jiang, Y., Qiu, L., Juarez-Perez, E.J., Ono, L.K., Hu, Z., Liu, Z., Wu, Z., Meng, L., Wang, Q., Qi, Y., Nature Energy, 4[7] 2019, 585-593. https://doi.org/10.1038/s41560-019-0406-2

[322] Billen, P., Leccisi, E., Dastidar, S., Li, S., Lobaton, L., Spatari, S., Fafarman, A.T., Fthenakis, V.M., Baxter, J.B., Energy, 166, 2019, 1089-1096. https://doi.org/10.1016/j.energy.2018.10.141

[323] Liu, D., Jing, H., Sa, R., Wu, K., New Journal of Chemistry, 43[24] 2019, 9453-9457. https://doi.org/10.1039/C9NJ01400D

[324] Chan, S.H., Wu, M.C., Lee, K.M., Chen, W.C., Lin, T.H., Su, W.F., Journal of Materials Chemistry A, 5[34] 2017, 18044-18052. https://doi.org/10.1039/C7TA05720B

[325] Boopathi, K.M., Karuppuswamy, P., Singh, A., Hanmandlu, C., Lin, L., Abbas, S.A., Chang, C.C., Wang, P.C., Li, G., Chu, C.W., Journal of Materials Chemistry

A, 5[39] 2017, 20843-20850. https://doi.org/10.1039/C7TA06679A

[326] Vashishtha, P., Metin, D.Z., Cryer, M.E., Chen, K., Hodgkiss, J.M., Gaston, N., Halpert, J.E., Chemistry of Materials, 30[9] 2018, 2973-2982. https://doi.org/10.1021/acs.chemmater.8b00421

[327] Perrakis, G., Kakavelakis, G., Kenanakis, G., Petridis, C., Stratakis, E., Kafesaki, M., Kymakis, E., Optics Express, 27[22] 2019, 31144-31163. https://doi.org/10.1364/OE.27.031144

[328] Öz, S., Burschka, J., Jung, E., Bhattacharjee, R., Fischer, T., Mettenbörger, A., Wang, H., Mathur, S., Nano Energy, 51, 2018, 632-638. https://doi.org/10.1016/j.nanoen.2018.07.005

[329] Juarez-Perez, E.J., Ono, L.K., Qi, Y., Journal of Materials Chemistry A, 7[28] 2019, 16912-16919. https://doi.org/10.1039/C9TA06058H

[330] Chaudhary, S., Mehra, R., International Journal of Engineering and Advanced Technology, 9[1] 2019, 4011-4016. https://doi.org/10.35940/ijeat.A1149.109119

[331] Li, J., Cao, H.L., Jiao, W.B., Wang, Q., Wei, M., Cantone, I., Lü, J., Abate, A., Nature Communications, 11[1] 2020, 310. https://doi.org/10.1038/s41467-019-13910-y

[332] Allegro, I., Brenner, P., Bar-On, O., Jakoby, M., Richards, B.S., Paetzold, U.W., Howard, I.A., Scheuer, J., Lemmer, U., Conference on Lasers and Electro-Optics Europe and European Quantum Electronics, 2019, 8871688.

[333] Masuda, T., Iyoda, M., Yasumatsu, Y., Sasaki, K., Zhang, Y., Ding, C., Liu, F., Shen, Q., Endo, M., Proceedings of SPIE, 10897, 2019, 1089710.

[334] Zhu, H., Fu, Y., Meng, F., Wu, X., Gong, Z., Ding, Q., Gustafsson, M.V., Trinh, M.T., Jin, S., Zhu, X.Y., Nature Materials, 14[6] 2015, 636-642. https://doi.org/10.1038/nmat4271

Keyword Index

s-orbital, 6, 16
spin-orbit coupling, 5, 12, 27, 36, 45, 54, 83
spiro-OMeTAD, 20-24, 33, 62, 68, 77, 87-88, 90, 92, 99
surface dipole, 20
symtriazine, 99

Tauc direct-transition, 82
tetragonal phase, 4, 58, 85
tetrathiafulvalene derivatives, 21

thermalization, 30, 66
thiophene, 92
titania scaffold, 74
toxicity, 5, 97-98
trihalides, 12
trion, 3
trivalent cations, 7

Wannier-Mott, 83

γ-butyrolactone, 19

About the Author

Dr. D.J. Fisher has wide knowledge and experience of the fields of engineering, metallurgy and solid-state physics, beginning with work at Rolls-Royce Aero Engines on turbine-blade research, related to the Concord supersonic passenger-aircraft project, which led to a BSc degree (1971) from the University of Wales. This was followed by theoretical and experimental work on the directional solidification of eutectic alloys having the ultimate aim of developing composite turbine blades. This work led to a doctoral degree (1978) from the Swiss Federal Institute of Technology (Lausanne). He then acted for many years as an editor of various academic journals, in particular *Defect and Diffusion Forum*. In recent years he has specialised in writing monographs which introduce readers to the most rapidly developing ideas in the fields of engineering, metallurgy and solid-state physics. His latest paper will appear shortly in *International Materials Reviews*, and he is co-author of the widely-cited student textbook, *Fundamentals of Solidification*.

www.ingramcontent.com/pod-product-compliance
Lightning Source LLC
Chambersburg PA
CBHW070736220326
41598CB00024BA/3443